The Political Mapping of Cyberspace

The Political Mapping of Cyberspace

Jeremy W. Crampton

The University of Chicago Press

The University of Chicago Press, Chicago 60637
Edinburgh University Press Ltd, Edinburgh

12 11 10 09 08 07 06 05 04 03   1 2 3 4 5

ISBN 0-226-11745-6 (cloth)
ISBN 0-226-11746-4 (paperback)

Cataloging-in-Publication data have been requested from the Library of Congress.

♾ The paper used in this publication meets the minimum requirements of the American National Standard for Information Sciences—Permanence of Paper for Printed Library Materials, ANSI Z39.48–1992.

Contents

Acknowledgements

I would like to thank the following people who provided input into this book, either by helping me think, or by providing material to think about: Joseph Pearson, Stuart Elden, Gunnar Olsson, Anne Knowles, Kylie Jarrett, Camille Duchêne, Margo Kleinfeld, John Krygier, Dona Stewart, David Weberman, Kathryn Kozaitis, Mei-Po Kwan, Nik Huffman, Scott Freundschuh, Julie Tuttle, and the contributors to the Foucault and Heidegger Spoon lists. In particular I have profited from a series of long-distance discussions with Stuart Elden. More locally the GSU Reading Group and the Seminar in Ethics and Politics of Mapping have been especially exciting and productive of both thought and practice, as has in a different way, the Java Monkey. Elaine Hallisey Hendrix provided vital technical advice. The Internet Archive (http:/webdev.archive.org) provided some historical web pages. Earlier versions of Chapters 2 and 3 were published in *Cartographic Perspectives* and profited from the suggestions of anonymous referees. Finally, I would like to thank my editor, John Davey, for his encouragement and support over a longer period than we had expected. Thanks.

Introduction

It seems to me that there was one element that was capable of
describing the history of thought: this was what one would call
the element of problems, or more precisely, problematizations.

Michel Foucault[1]

There have now been several rounds of commentary and analysis
about cyberspace, the Internet and the World Wide Web. It is
approaching a decade since the Internet reached wide consciousness in
the West and over thirty years since it was developed. While some scholars have shown great excitement over its potential for non-linearity and
hyper-thought, we have also seen an intellectual backlash against it as so
much silicon valley snake oil. Most recently we have seen the rise and fall
of its commercialization and dot-coms. If I were to order the books on my
shelves chronologically their titles would betray the increasing reality of
Internet studies: the gradual move from the Internet or information itself
as the unit of analysis ("The cult of information", "The virtual community") to its integration into studies of society on a wider basis ("The rise
of the network society", "Splintering urbanism"). This book continues
that trend by investigating the integration of cyberspace into everyday
life. It is a study of the *spatial politics of cyberspace*. The goal is to investigate how we find our place in the world. It is a spatial politics which takes
cyberspace as its domain.

PROBLEMATIZATION OF CYBERSPACE

Social life is inherently spatial. This does not only mean that our experiences "take place" in a spatial way, but that fundamentally we "are"

spatially. As existing beings we live in, open up, shape, and are shaped by spaces and places. We cannot be in the first place without being in space. If we grant this initial proposition three important consequences follow.[2] First, social space is often struggled over. The primary sense of struggle that I underline in this book is the struggle over eking out the possibilities of being spatially as far as cyberspace is concerned. That is, to enhance or augment the ways in which we are emplaced by making choices, assessing conditions, implementing changes, getting involved, putting projects into practice, and taking stands. It is usual to think of practices and engagements occurring *in place*; in that neither they nor we ourselves are abstracted from the world (e.g., see Casey, 1997). This raises the question therefore of how we are "in place" cyberspatially – is this phrase a metaphor or does it have a deeper meaning, perhaps even an ontological one about cyberspatial being?[3] Moreover, many emplaced activities are themselves concerned with the question of *what kind of places* we wish to live in. This question is often implemented as a struggle to realize a goal or vision. For example, how shall we run or govern our online community, and how shall it relate to other communities on or offline? Is online community engendered through communication, shared interests, or something else? Who shall be admitted to the community and on what grounds – that is, how shall we authenticate digital identity? What are the consequences of highlighting identity as the authentic? How shall we allocate communication technology resources across the city? Do we have an obligation to admit everyone? How do we make decisions politically that affect people in different places? Is cyberspace governed by the same laws of conduct as physical space (e.g., can a law in one place, such as France, be imposed on an American company's Web "presence" in France?).[4] And finally, what kinds of places and spaces do we ourselves construct (deliberately or not) in order to govern? These questions concern government, legality, spheres of influence, and the political constitution of cyberspace. As such, these spatialized struggles are inherently political questions. Thus our first consequence is that the political is always spatial.[5]

Second, many of these ways of living spatially depend upon and construct a whole variety of meanings and knowledges, only some of which are an issue for us at any given time. That is, while some conditions and spatial practices are accepted, from time to time other issues become sites of difficulties. When this happens we increase our concern toward those issues, make judgements and changes, and even potentially come to have a different set of possibilities for our lives. Just *why* difficulties emerge is a good question, but they clearly become so when oppositions are in play, when discourses and powers are opposed. Recent examples discussed in this book include the disputes over privacy and surveillance on the Internet, the role of digital technology in time–space compression and globaliza-tion, the construction of subjectivity as "dangerousness" in online crime-mapping, the spatial unevenness of the digital divide, and the formation or breakdown of communities as mediated by communication technology. This is not to say that problematizations can always be resolved, but that until something is taken up as a difficulty it cannot even be addressed. The second consequence therefore is that sites of problematizations are where a spatialized politics can have critical effects on our possibilities.

Third, a primary and important source of spatial knowledge is produced in mapping practices. Historians and anthropologists have now examined a wide set of cultural contexts in which mapping has been practiced and have documented a historical range extending into prehistory and across the vast majority of human cultures.[6] Maps preceded both language and counting. From the stick charts of the Marshall Islands, to the Aboriginal map-paintings of Australia (Watson, 1993) to the modern Western topo-graphic maps, the idea of the map is central to human activities (Harley, 1987a). Although the look, form and function of maps are incredibly rich and diverse, the history of cartography ably demonstrates that _maps provide a spatialized understanding of the world_ (this shall be our definition of mapping). Given the centrality of space articulated in the previous con-clusion this should not perhaps surprise us, although its full ramifications remain to be worked out (for example, are maps themselves a social space? If so, how might this challenge representational theories of mapping?).[7] A third consequence therefore is that mapping is necessarily part of our engagement with space and thus also part of our political concern.

In this book I shall use this interpretive framework to investigate the spatial politics of cyberspace. Cyberspace for us today has become the site of a number of problematizations and issues, ranging across debates over privacy and surveillance, access to technological resources and knowl-edge, the nature of relations between the physical and the virtual, iden-tity and authenticity, and the nature of spatial relations in cyberspace. At this historical moment cyberspace is an extraordinarily open site of con-testation. In the first chapter, the problematics of cyberspace will be dis-cussed as a Foucauldian *genealogy*, that is, a historicized enquiry into problematics at the current moment. How does cyberspace produce sub-jects through power relations? What are the relations between practices of the self and subjectifying power? How is confession constituted as part of the production of the truth? In effect my enquiry into cyberspace is a genealogy, partial and preliminary though it may be. And because cyber-space is (in part) non-material, maps of it have provided an incredibly rich imaginary of its nature, and have produced many of the ways we are accustomed to thinking about it (including the fact that it is a cyber-*space*).

The spatial politics of cyberspace has been significantly constituted in cartographic terms.

If, as I have argued, mapping is a necessary and critical component of politics, then it is also critical to know *what kind* of geographical knowledge maps produce, and equally what they could produce that they currently do not. In other words, how is geographic knowledge produced through mapping? Until the past couple of decades this question had not been posed as such in politics, geography, or cartography.[8] Starting in the 1980s, however, historians, geographers, philosophers, and sociologists began an analysis of the ways in which politics has intruded upon or imbued mapping so that maps serve particular interests. One of the commonest examples of this approach was to investigate propaganda maps, for example those produced during World War II by Germany (Guntrum, 1997). While many writers were prepared to accept that some maps had been used politically, only a few made the seemingly more radical case that _all_ maps are politicized and serve the interests of class and power. Those that did so, such as Harley (1988a, 1989) and Wood (1992), perceived their project as unmasking the "subliminal geometries" (Harley, 1988a, p. 289) of political ideology in mapping. But whether in the more limited case or the more general, the question of the relationship between mapping and politics was one which investigated how a taken-for-granted political arena played a role in maps and mapping. Black's recent history of maps and politics, for example, took as its project the fact that mapping "cannot be divorced from aspects of the politics of representation" (1997, p. 10).

But the investigation of the political usages of maps does not exhaust the relations between maps and politics. The question in this book is how maps and spatiality constitute the political for cyberspace, or, the political project of *how we find our place in the world*. In this sense, the map is not the target of the analysis, but the political, or more specifically how rethinking spatial representation can lead to a rethinking of the political. This distinction can be clarified with reference to a long-standing division in philosophy between knowledge pertaining to how things are, and the grounds or possibilities of having that knowledge. This distinction was powerfully analyzed by Heidegger in the Introduction to *Being and Time* (Heidegger, 1962) where he labeled these oppositions "ontic" and "ontological" knowledge respectively (§§3–4). The political analysis of maps has been one of ontic knowledge, the nature of things (e.g., how maps have a political encrustation or mask). If we proceed very far with this ontic analysis it soon becomes evident that mapping produces knowledge which has a particular conception of space as a container for social activity, a conception which causes a number of difficulties and political impasses.

A classic example is provided by redistricting, in which the results from the national census are used to redraw the borders of political districts using a very common form of spatial representation called a "choropleth" map. In a choropleth map an area or region is divided up into a number of mutually exclusive enumeration (counting) units and a single value (rate) of an socio-economic variable (population by poverty, race, or education for example) is assigned to cover that entire enumeration unit. Thus choropleth maps partition a continuous space into a number of, literally, "place-holders" into which a single spatially averaged number is allocated.

In redistricting, the debate over these districts revolves around the problems of making each district fairly "contain" the right mix of people in order to elect a representative, whose political decision-making then extends over the electoral district.[9] In effect, the party in power at the time of the census draws up these districts to maximize its likelihood of getting its members elected, a process which is divisive, contentious, and often resolved to no one's satisfaction in the courts (see Monmonier, 2001). An ontological counterpart to this ontic question might be to ask in what ways we can rethink mapping so that space is not characterized as a container, to rethink the production of identity by borders (or "bordering" – see Crampton, 1996), and the political effects of chopping up richly integrated places into enumeration counting units. What other spatial possibilities of being politically might be produced by different conceptions of space and different kinds of map? Clearly, remapping political districts produces and is produced by geographical knowledge, but it is equally clear that relatively little is understood about the ontological grounds for such ontic knowledge. Such a lack is even more keenly felt for a spatial politics of cyberspace.

But why cyberspace? Is this not a subject which has become emptied of meaning through multiple different assertions and difficulties? Haven't recent events shown that the Internet and Web bubbles have burst? Yes, this is precisely the situation and what at the same time introduces the very possibility for constructive change; for a new mapping of possibilities. It *is* the case that cyberspace is no longer new and exciting. The uncritical acceptance of anything and everything "cyber" is long since past. Nor is it any longer the source of constant capital accumulation or venture capital investment based on two-page business plans, nor does it have a clear relationship with society as the realm of new possibilities, nor is it necessarily an oppression.[10] Many writers remain enthusiastic about cyberspace, but there are many others adding their hesitant voices to the initial gold rush of enthusiasm. And what is cyberspace anyway? Although we speak "of" cyberspace as if it were a single object, we might also wonder if it is instead an unevenly distributed and complex set of interrelated technologies (the

Internet, the Web), practices (email, interactive mapping) which have multiple outcomes at different spatial scales (the digital divide, production of subjectivities). Entirely new problematics are beginning to emerge, such as the role the Internet plays in people's lives with regard to social engagement and alienation, or the fresh possibilities of self-expression and memoir writing known as "blogging".

The problematics of cyberspace can begin to sketch a critical history of its truth constituted as a problematics of thought. Because this is not a history of ideas or behaviors I will not therefore provide a chronology of the events and technologies associated with the Internet or the Web qua cyberspace.[11] My only excursus into the great imaginaries of cyberspace, for like reason, will be my discussion of the work of Philip K. Dick (Chapter 5).

Cyberspace for us now is a kind of mapping into which we project our fears and hopes: it is our "project". As such a projection, cyberspace presents a number of representational (and thus cartographic) issues. It is no accident that all maps are projections. The issues are not in mapping its abstractness (cartography has mapped the abstract for centuries) but with its spatial politics. Cyberspace is a classic case of a space which is produced, and which in turn produces (spatialized) subjectivity. Thus not only can cyberspace be mapped, but in tracing out its contours we are tracing out the lines on our own faces. Cyberspace is an area of geographic knowledge that sits equally between society and technology. It invites the "fieldwork" for a critical politics of spatial representation and from that a critical politics of geography and space.

OUTLINE OF THE BOOK

The questions this book addresses are grouped around multiple axes: the problematic of space and spatial representation pursued through the politics of spatial representation (Chapters 2 and 3); the problematic of subjectivity, analyzed through aspects of Foucault's technologies of the self (Chapters 4 and 5); the problematic of the production of cyberspace (disciplinary cyberspaces and the digital divide in Chapters 6 and 7); and finally the problematic of an ethics of the virtual, or the practice of cyberspace (Chapter 8). This ethics seeks the positivities of power and a pleasure of maps.

These axes are by no means exclusive, nor are they especially meant to indicate entirely separate fields. I have no doubt that they could be arranged differently or different aspects highlighted. More important is the project of a critical politics of cyberspace, and more specifically with the conditions of possibility for the production of geographic knowledge, a project which constitutes a genealogy. Prior to these discussions however, we need to return to and more fully clarify how a problematics of cyberspace is undertaken, and this shall form the subject of Chapter 1.

Part I of the book opens with a more detailed discussion on spatial representation by analyzing how one particular form of mapping, distributed interactive mapping, has risen to prominence historically in conjunction with digital communication technologies (Chapter 2). What effects are there on people's mapping experiences when map knowledges are no longer confined to the experts? How widely are online maps available and what functionality do they actually provide? I suggest a number of ways in which we might assess the criteria for answering these questions, and in Chapter 3 focus on the social production of geographical knowledge more directly (and the geography of access in Chapter 6). I do this by taking issue with the writing of Brian Harley to suggest that while he successfully initiated a project of politics and mapping, in many ways he remained mired in the modernist conception of maps as documents charged with "confessing" the truth of the landscape. I then discuss the possibilities of a non-confessional understanding of spatial representation and its implications for a politics of cyberspace. I suggest that instead of maps being interpreted as objects at a distance from the world, regarding that world with a view from nowhere, that they be understood as being in the world, as open to the disclosure of things. Thus a map is not a thing which represents (in the sense of re-presence). These possibilities are returned to throughout this book where I take up Heidegger's critique of the metaphysics of presence.

In Part II (Chapters 4 and 5) I discuss a double subject, that of authenticity and confession, by suggesting that cyberspace is undergoing a crisis of authentication. This identity crisis is well known, and, by some observers is celebrated for the way in which online identity is fragmented and destabilized (Turkle, 1995). Gender roles, for example, could be radically undermined and restructured to avoid the negative effects of normalizing imperatives. However, I would suggest two somewhat more worrying aspects of this crisis. First, problems of authentication have given rise to an ever-increasing and shifting set of truth-producing practices. How can we tell you're really who you say you are? These truth-producing responses, which I call a *confessional regime*, have the object of producing the truth of the subject by grounding authenticity in a stable subject. The cyberspace crisis in identity has therefore "called forth" just the kind of normalizing power relations which many people wanted to leave behind. Therefore I argue that the crisis of authenticity in cyberspace is constituted as a question of the essential truthfulness of the individual's identity. While better alternatives are not yet clear, one suggestion made here is that practices of the self such as *self-writing* may be able to dodge out from the authentication–confession binary. I therefore spend some time in looking at the recent phenomenon of "blogging" online (web journals or memoirs) and frank-speaking resistance (*parrhesia*) as just such practices of the self which do not seek to produce the truth of the self but to work on it as a project.

In Part III (Chapters 6 and 7) I discuss a pair of case studies of the production of the spatial politics in cyberspace. In the first, a look at computer crime-mapping, I suggest that subjectivity is spatially produced as an element of criminal "dangerousness". As with psychological profiling, computer crime-mapping works to produce "geo-profiling" of subjects and to make geographical predictions for culpability and guilt. Criminality has a geographical dimension that is produced by digital technologies. I look at the origins of crime-mapping and its emergence alongside the statistical governance of the state over its territory. This statistical rationality provides insight into the obsession today with at-risk human and natural resources, as well as providing the grounds for the surveillance explosion. In the second case study I look at the vexing issue of the spatial unevenness of access to communication technologies in the physical world. Here we see clearly that the digital is not divorced from the physical world, that cyberspace is not a free-floating realm where geographies do not matter. Indeed, these inequalities in access have continued to widen, which suggests that technology itself is not able to close the gap. Only better policies from the physical world of resource allocation, coupled with spatial policy-making tools which can be used at the local level by the people themselves, are likely to make inroads into this problematic.

No doubt problems of cyberspace do not necessarily have neat solutions. But they can be the stimulus and site of thinking and imagining. In my last chapter, I discuss the possibilities for an ethics of virtuality from the perspective of the spatial politics of practices. This ethics is not a morality in the sense of rights or wrongs, for that merely returns us to procedures of normalization and the production of confessionalized truths. Nor is it one which is totally divorced or ungrounded from power relations, for it does not see power as inherently negative but rather as productive.[12] It is an ethics as *ethos*. Although the discussion is grounded in theory there are a number of pragmatic avenues which offer themselves for exploration with the overall goal of maximizing freedoms of cyberspace. These include interpreting identity as non-confessional care of the self, of taking spatially embodied pleasures rather than highlighting desire, of maps as counter-memory, and of local community formation through spatial policy-making.

CHAPTER I

Being Virtually There: The Spatial Problematics of "Cyberspace"

Today we are faced with several critical problematics of cyberspace: its role in the rise of a global information society, and with it, increased surveillance and commodification;[1] worryingly dystopian futures if rampant commercialization and ecological degradation are not curbed (a vision informing the protests against the World Trade Organization (WTO)); the role of the local in the global and the loss of "authentic" places under globalization;[2] and finally the position of the subject in the era of virtuality and simulacra (Baudrillard, 1994; Egan, 2000).

Many, if not all, of these questions are often thought by political and economic theorists to be exacerbated by digital and voice telecommunications, the Internet, global flows of capital, deterritorialization, flexible accumulation and remote back-offices, and so on. Late capitalism then, would seem to have extended, co-opted, and reproduced through the channels of cyberspace. In some versions of this view, cyberspace seems to be merely a carbuncle on the face of society, with the implication that it is back in society where we need to perform our analysis (and cyberspace can be left to the engineers). In David Harvey's engaging book, he sometimes seems to imply this, writing for example that "it is easy to make too much of [the information revolution]", and that although "the newness of it all impresses" the "newness of the railroad and the telegraph, the automobile, the radio, and the telephone in their day impressed equally" (Harvey, 2000, p. 62). This view does two things. First, it affords primacy to the real physical world, as opposed to the simulated, virtual world. Second, it denies any vitality to the virtual, characterizing it as the mere product of the primary physical world, in fact perhaps even as a (special) subcomponent of the physical world: the ultimate spatial "fix" for an ultra-mobile capital. Any possibility of causality which emerges in the virtual and flows into the physical is not entertained. I argue that this view is mistaken.

On the other hand, commentators have often reversed cause and effect by identifying cyberspace as a major socio-cultural "turn" in the West (and perhaps later on the rest of the world via diffusion). These commentators tend to characterize cyberspace (either positively or negatively) as the source of a revolution in publishing (finally ending the domination of print), mapping/geographic information systems (GIS) (ending the domination of paper maps),[3] bringing in new threats or opportunities to privacy and morality requiring a juridical response, or freeing investment capital to circulate without restrictions of location or market hours (e.g., see Cairncross, 1997). A major subcomponent of this is a discourse of information and communication technologies as "enablers" of social and democratic development (DOI, 2001). For example, one study (Anderson et al., 1995) traced the linkage between email usage and level of democracy, with significant correlation between the two. (Although they drew no conclusions about cause and effect, the possibility of the communications technology providing what Habermas called the society of communicative action was there.)

Likewise, some cultural critics speak of new revolutions and "technology as transcendence" (Graham, 1998) of physical reality. For example, Paul Virilio sees cyberspace as a means to reach the last barrier, that of light, and move into a world of immediacy. We are now in "real time" he argues, which means the end of place: "having attained this absolute speed, we face the prospect in the twenty-first century of the invention of a perspective based on real time, replacing the spatial perspective" and "it poses a threat to geopolitics and geostrategy" (Virilio, 1995, p. 2). Cyberspace is unprecedented, he enthuses, with the effect that "perceived reality is being split into the real and the virtual . . . to be is to be *in situ*, here and now, *hic et nunc*. But cyberspace and instantaneous, globalized information are throwing all that into total confusion . . ." (p. 2). Cyberspace is a disemplacement.[4]

As one character says in the film *eXistenZ*:

> There's reality, there's what you see on TV, and there's virtual reality. Guess which one is going to win the day? . . . reality in all its forms is being threatened now, more than ever. It is being eroded and it is washing away in the deforming storm of nonreality, which masquerades as reality and which will eventually replace it if we do not take the appropriate steps.
>
> (Priest, 1999, pp. 105, 190)

In this book I examine whether these two positions tell the whole story. I seek to critically *problematize* these unexamined discourses and their cartographics.

The emergence of a critical theory and politics of cartography and mapping over the course of the 1990s has a number of far-reaching implications, but, I would suggest, has only been addressed in a number of specific localized debates.[5] Cartography and GIS are often thought of as central to the disciplines of geography, urban planning, locational based services (LBS), economic and political geography, etc., but are very infrequently theorized with any panache. This is all the more perplexing when critical geographers often remark in passing that "of course" the standard account of mapping as a neutral reflection of the landscape is not tenable.[6] With regard to cyberspace, the first question that geographers have tended to ask is "how can we map cyberspace?" or "what are the geographies of cyberspace?" (e.g., see Dodge and Kitchin, 2001; Crang, Crang and May, 1999). These cartographic questions have been cast without a ready-to-hand theorization of mapping. As I have noted elsewhere (e.g., Crampton, 2001a) this is understandable (to a degree) given the only rather shaky attempts at theorization from within cartography and GIS itself. Cartographers and GISers themselves in fact see no such necessity to problematize mapping and often proceed with a standard "maps as communication devices" model of cartography (for a recent example, see Goodchild, 2000). The first task therefore is to consider what a theory of cartography – call it a critical cartography – might look like, and in the first two chapters of this book I attempt to provide a few sketches towards a theory of cartographic power-knowledges.

Secondly, we might consider the question of how cyberspace itself is defined. By delineating a separate sphere called "cyberspace" we run into trouble since it objectifies a realm of meaningful activity which I shall argue is folded (and manifolded) in the physical as part of our world. Furthermore such objectification tends to indicate a unified object rather than a rich field of multiplicities and inequalities. In this book, I shall try to avoid the common approach of characterizing cyberspace as a singular thing or place, a separate entity.[7] There is often a two-fold maneuver when cyberspace is objectivized: first, as we might expect, it is privileged as a separate domain, as an *object* of analysis which is knowable as a thing, and second, it is cast as artificial, or virtual. These maneuvers result in a strange tension, where cyberspace is simultaneously privileged and de-privileged. For example, Benedikt defines cyberspace as "a globally networked, computer-sustained, computer-accessed, and computer-generated, multidimensional, *artificial* or '*virtual*' reality" (Benedikt, 1991, p. 122, emphasis added). The real is opposed to (and being replaced by) the "deforming storm" of the virtual. Implicit in this idea is a de-spatialization and disemplacement – "real time [inaugurated by cyberspace] now takes precedence over real space" (Virilio, 1995, p. 2).

By resisting these maneuvers I suggest that an objectified cyberspace is the wrong level of analysis. The way cyberspace is treated here is not transcendentally (that is, as some "thing" beyond us that is "less" real, or, as for *eXistenZ*, a replacement of the real) but as a *mutual* process of production between physical space and abstract or virtual space, as a series of relations, and as a process of becoming. We might call this mutuality "being virtually there".

My questions here then are not "where is cyberspace?" or "what are the freedoms of cyberspace?" but rather "what are the material relations of the production of cyberspace?" and "how is cyberspace extended by the relations of power-knowledge?" and "how is the subject formed in relation to cyberspace?" (how do we have our being in cyberspace?). In other words, I am interested in the condition of possibility for cyberspace. In doing so, I employ a fairly generous understanding of cyberspace, which is necessary in order to fully grasp the interrelations between physical and virtual space. This approach we can call, after Foucault, one of "problematization", where we see "the development of domains, acts, practices, and thoughts that seem . . . to pose problems for politics" (Foucault, 1997a, p. 114).[8]

By thus emphasizing such a *problematics* of cyberspace, I wish to signal the fact that cyberspace is an ongoing outcome of the material and virtual relations of production, that the subject finds him or herself a part of this production, and is in turn produced. But it is the producing which is important, not cyberspace as an end-product.[9]

But *problēma* is also a shield,[10] and, furthermore, a bulwark and riddle, from the word *proballō* for anything thrown or projecting (Liddell and Scott, 1990; Crane, 2001).[11] Heidegger has sensitized us to some of the meaning of "projection", notably as a reaching out to our (future) possibilities. So to problematize is to take something up as a difficult but essential opening up of human possibilities. This is the sense I want to underline with respect to cyberspace. Socrates threatens to toss out bits of knowledge to the ignorant Strepsiades in *Clouds* (line 490), like a dog being tossed scraps.[12]

More explicitly, by "problematics" of cyberspace I mean the following:

1. It is to set something as a question, either by us, or more usually one which occurs in the discourses of the time and place under study. It is to throw something forward or set it out as an issue. Why and how is cyberspace at issue?

2. Second, to problematize something is to undertake a history of thought, rather than a history of ideas or representations. Histories of cyberspace (e.g., Hafner and Lyon, 1996) tend to follow the progressivist

chronological emergence of technical developments rather than <u>how</u> <u>our relationship to technology has opened up or closed off</u> meaningful <u>thought and practice</u>. An illustrative corrective is offered by Foucault, namely that "it was a matter of analyzing, not behaviors or ideas, nor societies and their 'ideologies', but the *problematizations* through which being offers itself to be, necessarily, thought – and the *practices* on the basis of which these problematizations are formed" (Foucault, 1985, p. 11, original emphasis).[13]

3. Finally, to problematize is to examine the truth claims of the discourses: "problematization doesn't mean representation of a preexisting object, nor the creation by discourse of an object that doesn't exist. It is the totality of discursive or non-discursive practices that introduces something into the play of true and false and constitutes it as an object for thought" (Foucault, 1988c, p. 257).

Foucault achieved his problematization by historicizing his subject matter. With these three aspects of problematization in mind, I would like to attempt, if only in a preliminary way, a *spatial* problematization of cyberspace, or what I label "being virtually there". The advantage of this label is that it invokes being, presence-absence, digitality, spatiality, and amalgams of these terms. Table 1.1 attempts to interpret these terms and to assign them a place.

The manner in which I wish to do so, admittedly only performing a partial analysis, has three components: a power-knowledge of representation understood through mapping; the technologies of the self constituted through and via cyberspace; and the material relations of production, again, generally through and proceeding from cyberspace. In sum, this means to put into play the question of the discourses and institutions of cyberspace.

Table 1.1 The problematics of the book: a genealogy of the subject for cyberspace

Term	Topic	Enquiry
Being	Problematization	Problematization (1)
		Authentication (4)
Being virtually	Digital being, being present-absent	Confession and Parrhesia (5)
		Disciplinary cyberspaces (6)
Virtually	Almost, deferred, digital	Geography of the digital divide (7)
Virtually there	Digital spatiality	History of cyber mapping (2)
		Why Mapping is Political (3)
Being virtually there	Ethos	An ethics (8)

THE PRODUCTION OF CYBERSPACE

The production of space is best known through the work of Henri Lefebvre (1991) who argued that space and spatial relations are the material and social outcome of capital. The totality of relations and practices between things and people, the discourses or representations of space, all act to produce space. They do not just act within a passive space, but actively generate or produce real spaces and places. The subsequent explication of the production of cyberspace has been usefully initiated (Bukatman, 1993; Graham, 1998, Dodge and Kitchin, 2001), but the topic has remained relatively unexamined compared to the quantity of literature we find on the production of space more generally (e.g., see Brenner and Elden, 2001). Why is this? The question now seems pressing, given the urgency of the problems outlined above: how, and with what effects, is cyberspace produced?[14]

As a powerful example of this point, consider gentrification, which is rapidly occurring in metropolitan areas across the country, especially in inner working-class neighborhoods which are now predominantly African-American. In Atlanta, for example, "gentrifiers" (a broad group including prospective and recent home buyers, small renovation companies and real estate developers) often exchange knowledge and organize community meetings via email, not as a deliberate exclusionary tactic, but because it is convenient. At these meetings they organize neighborhood watch groups, decide on rules for behavior and other tools for management of the neighborhood (speed limits or speed bumps, routes of school buses, etc). Since the long-time local residents are too poor to own computers or use email they are excluded from this realm of activity and decision-making. A neighborhood can therefore be socially constructed through differential access to knowledge. Certain spatial relationships can also be created between different sets of neighbors, differentiated clientele in neighborhood bars and restaurants and so on.[15]

Various attempts have been made to understand the complex set of relations between place, space, virtuality, and physicality. Broadness may allow us in an initial analysis to consider a range of practices without unnecessarily confining the argument. In this way a number of loosely conjoined topics can be discussed and their specifics can be brought to bear in the particular chapters. It is probably sufficient then to say for now that our topic is spatiality in a digital and virtual context, including computer situated representations of place and space such as computer crime-mapping, and how cyberplace and cyberspace are constructed and problematized.

Virilio thought that the inauguration of a terrain of instantaneousness meant big problems for political democracy, because democracy was tied to the local level of place and cities. But as I shall discuss in this book, the problematic is far more fundamental: precisely because cyberspace is *not* new, because it is *not* split from the real, and because it is *not* just objectively present, the problem for politics is at its greatest. As long as cyberspace is a problem, we will always have something to do.

SUBJECTIFICATION AND CYBERSPACE

If characterization of the terrain of cyberspace is problematic, then no less so is that of the subject or agent who interacts with it. We should be fairly familiar by now with analyses of the psychology of cyberspace, of how people's attitudes vary in relation to it (do they want it, fear it, get more divorced from society as a result of it, etc.). There have been a number of reports (and rebuttals or dismissals) of how the Internet is alienating people from society, in language quite reminiscent of that used to describe video games in the 1980s.[16] My questions here are again somewhat different because I wish to focus not on the attitudes of people "in" cyberspace but how the subject is produced in relation to cyberspace, a process we can call subjectification (see box).

Subjectification should not be thought of as a passive event. Cyberspace, as a particular domain of the world, does not cause waiting subjects to be in one way or another. However, it is part of our world even if we reject it or turn away from it. What are the strategies by which the subject forms him or herself (what Foucault called the "technologies of the self")? As Elden (2001a, p. 109) has pointed out, for Foucault *technê* is understood as a practice and is not limited to the modern sense of the word "technology". In Greek, *technē* very often has the sense of producing or the production of something which was not previously present (Sallis, 1999, p. 15). The "technology" of the self therefore is concerned with the production of the self. It is to these issues that I wish to turn in Part II of the book (see Chapter 4).

GOVERNMENTALITY AS THE "CONTACT POINT"

It should be no surprise that in the information era we are concerned with the circulation of information; how much information circulates, where it does so, who has access to it, what constricts it? Fifteen years ago Brand noted the basic equation: "information wants to be free, information wants to be expensive [controlled]" (Brand, 1987, p. 202). During the evolution of the Web in the second half of the 1990s one could often see

Subjectification

Subjectification (or the genealogy of the modern subject) was of long-standing interest to Foucault, and as is well known, he engaged with it in a number of ways over the course of his career. A flavor of his interest can be gained from some lectures given in 1973 in Brazil on the subject of truth and its juridical forms (Foucault, 2000a). These lectures delineate a "history of truth" and they reflect many of the themes of his book *Discipline and Punish* (1975) as well as Volume I of the *History of Sexuality* (1976). Foucault was keen to let his audience know that truth was not out there, waiting to be uncovered. Instead we should understand that truth is produced; not made up exactly, for it has its causes, but at least constructed. From whence then comes this truth? From political power-knowledge relations. What are its effects? The construction of the subject. "There cannot be particular types of subjects of knowledge, orders of truth, or domains of knowledge", thunders Foucault, "except on the basis of political conditions that are the very ground on which the subject, the domains of knowledge, and relations with truth are formed" (Foucault, 2000a, p. 15). Following Nietzsche, Foucault saw knowledge in an adversarial struggle with the world: "there can only be a relation of violence, domination, power, and force, a relation of violation" (p. 9). Knowledge then is born of struggle and power.

this debate in different contexts: privacy vs. access, intellectual property and copyright (e.g., in the wake of the Digital Millennium Copyright Act in 1998), attempts to legislate decency online (e.g., the unconstitutional Computer Decency Act (CDA) and its sequel (the Child Online Decency Act or COPA, passed in 1998 but currently under injunction). COPA (*Ashcroft v. ACLU*) has the added dimension that it attempts to regulate cyberspatial speech according to "community" standards. These standards are not derived from where the speaker lives or is physically located, but potentially from *any* community (the expectation is that it would be the most conservative community). A legislator in such a community could "reach out" to a website anywhere else and indict its owners. Legislatively speaking therefore, this law (should it pass constitutional muster) pits our notion of community derived from our bodily location against our cyberspatial "presence". Is the very notion of community standards obsolete?[17]

Privacy and data access alone have constituted huge areas of concern as people have struggled with the notion of privacy erosion caused, it would seem, by the increasing availability of personal data on the Internet, the

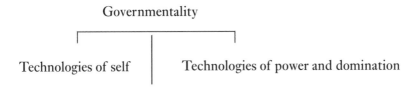

Figure 1.1   The "contact point" (interface) of cyberspatial governmentality.

2000 census in the USA (which featured Republican members of Congress urging people not to complete the forms), and corporate web-sites that track behavior down to the micro-level (e.g., how long you dwell on a particular web page). On the other hand there are a whole series of resistances to what might be called the globalization of corporatized infor-mation, attempts to interfere in the circulation of information (such as the Culture Jammers group, or thetruth.com; a group funded out of the tobacco agreement in the United States that has attempted to intervene in the circulation of smoking information) and resistance to the production of global brands and the co-option of community by corporations (Klein, 2000). There are also concerns that the data glut is leading to a kind of dis-affection, a cognitive blankness and passivity in the face of mountains of data which will decrease the workings of democracy and reduce voter par-ticipation levels. Others are concerned that our society suffers from "data smog" (Shenk, 1998) or too much spam (e.g., the Coalition Against Unsolicited Commercial Email, CAUCE), while yet others develop tech-niques for more efficient and powerful search engines to find useful infor-mation nuggets. Information wants to circulate freely, and information wants to be constrained. Nothing too new perhaps: "the association of a prohibition and a strong injunction to speak is a constant feature of our culture" (Foucault, 1997h, p. 224).

How can we navigate through these opposed prohibitions and orders? One suggestion is to reconceptualize these issues as a "contact point" between technologies of domination and technologies of the self (Foucault, 1999, p. 162). This contact point is where two forms of "government" come together, the government of the self by the self (or at least attempts to do so), and the activities more generally understood by the word govern-ment, i.e., the regulation of groups of people for the purpose of society and institutions. Therefore we can call these two spheres "governmentality" (see Figure 1.1).

This idea of the contact point is a useful one, therefore, because it allows us to move back and forth across different scales of analysis, from the individual subject to the wider context. The contact point snaps the binary division between a whole range of relations such as public–private, local–global, free and constrained information, and physical–virtual. To replace them are a whole series of strategies between which it is possible to move back and forth in a highly productive manner, for power is productive, and these relations produce regimes of power-knowledge.

CONFESSION AND *PARRHESIA*

Part II of the book lays out what I see as two competing manners in which the subject is formed in cyberspace. These are complex and deeply interwoven topics. Nevertheless I have divided them by opposing on the one hand the creation of an "authentic" self through strategies that construct a stable identity which one is enjoined to confess, and on the other a notion of authenticity as achieving our own possibilities through self-technologies such as self-writing, speaking truth to power (*parrhesia*) and blogging. Digitally speaking, when authenticity is reduced to self-*authentication*, we have always already narrowed the set of possibilities open to us. "Authenticity" today has been confined to just this notion of authentication of the self, in which confessional strategies act to produce the truth about oneself. An example of these strategies is the password log-on that many of us perform everyday. If we now see this as the construction of a stable and normalized identity, this is very far from authentically working on ourselves. In the two chapters in this part I discuss some possible technologies of the self that I think have recently emerged in cyberspace. The aim here is not to identify these particular strategies (self-writing, blogging and *parrhesia* as resistance) as the goal for anyone interested in a richer sense of self in cyberspace, but rather to examine how they have widened the set of possibilities. Strategies change and are replaced with others (already blogging is becoming splintered between different constituencies such as war blogging, news blogs, and journalism blogging).

CASE STUDIES IN THE PRODUCTION OF CYBERSPACE

> The society of tomorrow will splinter into two opposing camps: those who live to the beat of real time of the global city, within the virtual community of the "haves", and the "have-nots" who survive in the margins of the real space of local cities, even more abandoned than those living today in the suburban wastelands of the Third World.
>
> Paul Virilio (1997, p. 74)

With these gloomy words Virilio presents us with his vision of the future. In this vision we are treated to a model of technology driving and diffusing out over society for some, while others are left out in the backwash. To some degree, this vision is already here. In Part III of the book, I want to examine some of the ways we are constituted in cyberspace. We have been hearing for some time about the collapse of the spatial, whether it be the "global delocalization" of Virilio, the "death of distance" (Cairncross, 1987; Wilson et al., 2001), William Mitchell's "dematerialization" and spatial loosening, globalization in various guises, and so on. No doubt these are already looking a bit jaded and that's why a phenomenon such as Hardt and Negri's *Empire* can catch the imagination with its talk not just of deterritorialization but of *re*-territorialization, a respacing, which opens up the question of the local and the global all over again. Perhaps too we need to revisit the question of the local in the global for purposes of *resistance*, of space-time compression. In Chapter 6 I look at the use of cyberspace in (re)producing a disciplinary space of spatial criminality, geo-profiling, and crime-mapping. These practices mesh and merge the physical and the virtual in a discourse of "public safety", "dangerousness", and above all our "where". Where is the criminal near to, where does the criminal go, and ultimately how can we control the criminal's location. Crime-mapping activities which involve the digital and virtual provide an excellent case study of a new disciplinary society.

MIND THE GAP

In Chapter 7 I argue that the "digital divide" has different characteristics at different spatial scales. At present, very little work has been done at the local scale. One reason for this is availability of reliable or comprehensive statistics. At the international scale it is evident that the digital information society has made only a very limited and highly concentrated impact. Something like 91 percent of the world do not have Internet access (as of year end 2002). Internet diffusion is still almost completely concentrated into a few, rich countries, with the rest of the world digitally divorced. At this scale the important factors are telecommunications policy, income, and the barrier of international debt and literacy.

But even within the richer, more developed countries access is quite surprisingly variable. In fact, it is not so much the haves and the have–nots but a more continuous variability in access, or digital equity. A non-binary division is politically more useful because "divide" suggests that with sufficient investment and the right kind of programs the gap can be identified, *known*, and overcome. You could call this "minding the gap" in the sense that the problem is apprehensible and therefore solvable. It is often suggested, for example, that eventually cheaper computers and bigger bandwidths will erase the divide. But things are more complex than this technological fix suggests. We should rather think about successive rounds of inequality like waves on the beach; by the time one wave races up the shore (e.g., Pentium computers, landline telephones) soaking into the sand, the next wave has always already begun (e.g., Pentium II computers, Web-enabled cell phones, broadband) but it *lags* behind. And of course there are those at the top of the beach, beyond the high water mark who are forever left "high and dry", where no technology ever can reach. A digital lag is not a perfect metaphor because it suggests that given enough time things will be caught up, but it is successful at getting across the idea that difference is permanent. A gap or divide is knowable and thus fixable because you can compare what someone has on the one hand, or what someone else does not have on the other, or you can compare what someone has to some sort of ideal. Therefore the solution would appear to be: give the person who is going without the computer/cell phone/Internet portal. Once this is done we move on; no more gap, we've built a bridge.

Derrida (1994) offers an interesting perspective on the question of the ideal. In talking about the overall set of problems with the world we can delineate at least two ways to interpret what is going on. In the first interpretation we acknowledge the problems, but "the value and the obviousness of the ideal would not be compromised, intrinsically, by the historical inadequation of historical realities" (p. 86). That is to say, it is the gap between reality and ideal that is of moment, but just because reality falls short does not mean that the ideal is compromised. Our task, in this interpretation, will "remain indefinitely necessary in order to denounce and reduce the gap *as much as possible*" (p. 86, original emphasis). Here the gap is "minded" – it is known, quantized, and strategically targeted. As the gap is overcome, we can call it "progress;" we are moving from the empirical reality towards the ideal in ever increasing strides.

But there is a second interpretation, inherently in competition and incompatible with the first. Beyond the gap there is the "question of putting into question again, in certain of its essential predicates, the very concept of the said ideal" (p. 87). Derrida seeks to problematize the idea of the ideal itself.

So when we ask "who has less than ideal access to a computer" we should also be asking what is the ideal in the first place, and how is it determined? For example, consider this statement: "By the fall of 2000, almost all public schools in the United States had access to the Internet: 98 percent were connected" (United States Department of Education, 2001). But what does this mean in practice? Maybe there's a wire that has been pulled through, but where does it go: the administrator's office or the computer lab (if there is one)? And if it goes to the lab, does this mean that Internet-capable computers are connected to it or old PCs and Macintoshes? And if they are new, does this mean that a teacher is available who knows how to use them? And if there is, is the teacher full-time, is there space in the curriculum for a critical engagement with the technology, are there support personnel for the equipment, how often can the class use the equipment? In a study at Vanderbilt University, it was found that only 29 percent of black households had a computer compared to 44.2 percent of white households (Hoffman and Novak, 1998). The study also found that at each education level, whites were still more likely to own a home computer than blacks. When income is controlled for (i.e., compare whites and blacks at the same income levels) whites were still much more likely to own a home computer. This last finding of course is especially striking because income and education are often assumed to lie behind the digital divide. The Vanderbilt study suggests that, even for people of similar income and education, race is a factor.

In the fourth in a series of national surveys, the US Department of Commerce examined "digital inclusion" across a variety of variables, including race. They found that black households were actually *falling behind* the national level of Internet access over time. The gap of 18 percent between national and black Internet access in 2000 was wider by 3 percent than the gap in 1998 (NTIA and ESA, 2000). They also found that race and education only explain *half* the differences between black and white Internet access.

These are critical issues: is the divide narrowing or widening? Why does race seem to play a role? If access is diffusing unevenly, then *where* are the needs greatest: in urban, suburban, or rural areas? At the urban level, is it possible to build an "access surface" map that will reveal the white holes where people are "falling through the net"? Beware though of only thinking of this issue as one of technology, rather than how we *use* the technology in the modern information economy. Obviously technology is easier to measure. But as I suggested above, mere access to technology does not imply that critical thinking, rich experience, or meaningful participation are occurring.

TOWARDS A CRITICAL POLITICS OF THE PRACTICE OF MAPPING

As I briefly mentioned in the Introduction, there is a strand of thought in philosophy which makes a critical distinction between knowledge about things and knowledge of the basis upon which we know things, or more simply what we know and how we know what we know. In Heidegger this distinction emerges as the difference between *ontic* knowledge, which is "knowledge pertaining to the distinctive nature of beings as such, it is the knowledge of the sciences", and *ontological* knowledge which is knowledge of "the basis on which any such theory (of *ontic* knowledge) could be constructed, the *a priori* conditions for the possibilities of such sciences" (Elden, 2001a, p. 9). Heidegger expressed these possibilities over a historical horizon.

I noted above that this was a *critical* distinction and certainly there is every reason to believe that it should inform a critical approach. For Foucault, an analysis of the history of thought became possible at moments of problematization, when things became a problem or difficulty. In the last chapter of this book I return to this question explicitly. If Chapters 2–7 mark particular moments in the spatial problematics of cyberspace, Chapter 8 puts back into the foreground the ongoing project for which this book merely provides a ground-clearing: a critical politics of mapping.

The larger question of "putting into practice" is also discussed in this last chapter. I consider mapping to be a political practice and therefore it both benefits from the kind of relevance which we are seeking but also suffers from an already existing politicization. It was in order to avoid this trap that Foucault spoke of focusing not on desire (it was already politicized in that it ordered people according to desire as, for example, gay or straight) but rather on pleasure (Foucault, 1978). There is a chance, Foucault says, that because pleasure has not yet been politicized that we can partake in practices as a kind of exercise of freedom, as an ethics or *ethos* (McWhorter, 1999). The pleasure of mapping as a practice (*ethos*) is explored in this chapter through an examination of a kind of fetish-text: a short essay by Harley on the relationship between the personal and mapping.

CONCLUSION

The overall goal of this book is to attempt an opening up of the possibilities for a critical politics of space through Foucault's idea of problematization. I argue that taken as a whole, "cyberspace" constitutes for us now just such a political problem of space. The middle chapters investigate particular specifics or moments in this spatial problematization of cyberspace, including the production of subjectivity, representational configurations of power-knowledge, and the production of cyberspace.

Is this book then a Foucauldian analysis of cyberspace? Say rather that it adopts tools and ways of thinking that are offered by Foucault where useful, but it is not an attempt to "do" Foucault in cyberspace. I believe Foucault offers us a vocabulary with which to talk about cyberspace, but we are not confined solely to that vocabulary, nor really is there a single vocabulary to be aimed for. There is, however, a useful set of tools for understanding how space is ordered through power relations in Foucault's work which has so far not been picked up. If writers are useful, they are useful in the same sense that one uses a spade in the garden, and like a spade they can be easily discarded when it rains, only to be taken up again later in another way. I have little sympathy with some commentators who think that treating Foucault as a toolbox is somehow misguided, in that it will miss the true essence of his work. This seems unnecessarily foundationalist and unproductive – we need to explore his provocative writings for ourselves as a practice or experience, rather than a set of doctrines.

PART I

Cartographic Power-Knowledges

CHAPTER 2

The History of Internet Mapping

In what ways have maps and GIS operated to produce truth in cyberspace? Is it possible to give a non-progressivist history of how online distributed mapping has been implemented and deployed? What cartographic practices have been an actual part of cyberspace? In this chapter I focus on the history of the deployment of Internet mapping and the ways in which it acts to produce cyberspace.

The topic of this chapter is an important new development in cartography, GIS, and geography, which may prove to fundamentally change the way that spatial data is accessed, analyzed and communicated. With the explosion of the Internet and its convergence with geographical tools, it has become increasingly feasible to make spatial data display and analysis available to a wide, asynchronous audience. Variously labeled "Internet GIS" (Peng, 1999), "GIS Online" (a regular column in trade journal *GIS World*), and "Web-based GIS", in this chapter I use the term "distributed mapping" because "distributed" captures an important aspect of the activity: its highly dispersed, multi-user nature. The term is also relatively old, dating back to the 1970s in the context of distributed databases.

Much distributed mapping currently occurs on the Internet or the World Wide Web, but it can occur elsewhere too (and historically did so). For example, a distributed mapping environment could be made available via an intranet, or a hybrid CD-ROM/network product. The term "mapping" is preferable to another suggested term, "distributed geographic information" (DGI – see Plewe, 1997) because the latter can include the distribution of non-interactive spatial databases, and is also associated with a particular technology (GIS). Distributed mapping is not a technology, but a strategy. In addition, I emphasize the creative problem-solving and visualization capabilities of mapping (i.e., to see it as an interactive process of spatial knowledge discovery and creation).

Mapping as a spatial problem-solving activity is likely to endure whatever the technological manifestation, though its specific problems may differ.

Although distributed mapping is recent, it has risen rapidly, a rise that parallels the growth of the Web itself. However, it is as yet little understood; the implications and research issues (e.g., on map design, on geographic education, or on how space is represented) are not yet fully identified, let alone solved. One way to increase understanding is to examine the way distributed mapping is historically related to developments in cartography, GIS, and geography, as well as to larger societal developments such as the Internet.

DEFINITION OF DISTRIBUTED MAPPING AND SCOPE OF CHAPTER

What is distributed mapping? The critical concepts are:

1. access to spatial data processing and visualization tools to a dispersed audience;
2. interactivity with map or a spatial database;
3. spatial problem-solving or visualization need.

A "distributed" system is one which has elements of dispersion (L. *dispargere* to strew) and dispensing (L. *dispendere* to weigh out). In distributed mapping maps are therefore spread out (dispersed) but also (inter)actively allotted on demand (dispensed).

A typical implementation of a distributed mapping system would comprise a spatial data server, a network, and access via client computers (Figure 2.1).

This is the simplest and most inclusive model – there are many variations in practice (Plewe, 1997), some of which are discussed in this chapter. In the simple scheme illustrated the Internet or the Web can comprise the network. The two are not identical. The Internet was developed during the 1960s and the Web in the 1990s as a more user-friendly interface to parts of the Internet (Hafner and Lyon, 1996); however, both employ TCP/IP (Transmission Control Protocol/Internet Protocol) to send packets of data across networks. In Figure 2.1 spatial data is served out across the network and is interactively accessed by multiple clients. The server implementation can vary; it can include a separate HTTP (hypertext transfer protocol) server and map/GIS server, or these may be combined. Where most of the processing is done by the client the term "thick client" is sometimes employed; where the server assumes the bulk of processing the client may be a "thin" one (Peng, 1999).

SERVER———————————————NETWORK————————————————CLIENT(S)

Figure 2.1   Idealized schema for distributed mapping.

An interesting variation on this scheme is provided by Public Participation GIS (PPGIS), an outcome of the Society and GIS Initiative 19 (Pickles, 1999; Craig, Harris, and Weiner, 2002) in the National Center on Geographic Information and Analysis (NCGIA). The goal of PPGIS is to provide access to the full functionality and data of a GIS at the local level without necessarily employing a network. The GIS may be in a mobile van which visits different neighborhoods, or returns to the same neighborhood over time. Because mapping capabilities are being distributed to a wide and multiple audience who interact with the data (e.g., in making local planning decisions during road construction) it is appropriate to call PPGIS "distributed mapping".

This chapter covers distributed mapping, with an emphasis on interactive systems which provide massively distributed but individually tailored maps. It is not a history of digital cartography as a whole. Obviously there are overlaps with related developments such as mapping software and the history of GIS, but I do not consider these here.

CRITICAL THEORETICAL ISSUES OF DISTRIBUTED MAPPING

A number of theoretical issues have been raised in the literature which impact the task of anyone attempting to offer a history of problematizations of cartographic practice. These theoretical issues include: the relationship of maps and power; what a good theory of representation in cartography would be (e.g., empiricist or constructivist); how maps and mapping practices develop (their history) especially in regard to the notion of contingency; and how spaces and places are produced in mapping environments.

There are a number of possible responses to these problematics, and cartographers and geographers have at one time or another adopted them all:

1. The issues are irrelevant, are not accepted, and need not be engaged (theory avoiding).
2. The issues are already known, have been accepted, and need not be further engaged (theory embracing).
3. The issues are important, are still unresolved, and need to be engaged (theory engaging).

The response to these issues marks out what kind of cartographer/geographer you are. If you believe that (1) "The issues are irrelevant, are not accepted, and need not be addressed at all", then theory simply gets in the way of the job. When Derek Gregory says that "advances in GIS . . . assume that it is technically possible to hold up a mirror to the world and have direct and unproblematic access to 'reality'" (Gregory, 1994, p. 68), for example, is this something you could imagine discussing with your students or repeating to your boss? If not, then you are a Theory Avoider.

If you are a member of the second constituency, the ones who respond with (2) "The issues are already known, have been accepted, and need not be further addressed", you may recognize in Gregory's remarks an attack on the correspondence theory of representation implicit in cartographic practice for most of the second part of the twentieth century. Correspondence theory is the idea that a neutral, objective representation of reality can be made in maps, language, or other sign system, and that it is our goal as cartographers to do so. You may also feel that this critique is already being successfully mounted against cartography via the work of, *inter alia*, Harley, Wood, Edney, or Pickles. You are a Theory Embracer.

What is interesting is that members of Group 1 and Group 2 seem to face in opposite directions like some modern Janus of cartography. For every worker in an intellectual environment in which theory is only a distraction from the problem-solving capabilities of GIS, another scoffs at the idea that cartographers still employ the map communication model. For the most part, these constituencies have occupied different realms of discourse, and although various attempts (e.g., Pickles, 1995, 1997; Wright et al., 1997) have been made during the 1990s to bring them together, these have to date been occasions for the re-establishing of prior convictions, rather than intellectual movement. To clear decks and define terms is an important step in critical engagement, but as yet the fray has not been enjoined in a wider sense.

Members of the third group who respond "The issues are important, are still unresolved, and need to be addressed" may be forgiven for having a sneaking admiration for Eagleton's adage that "hostility to theory usually means an opposition to other people's theories and an oblivion to one's own" (Eagleton, 1983, p. viii). Theory Engagers believe that the other two groups lack critical engagement with theory, Group 1 because it prefers to *ignore* theory and Group 2 because it seems too *entranced* by it. Members of Group 3 want to argue with the technological determinism of Neil Smith's assertion that the "Gulf War was the first full-scale GIS war" (Smith, 1992) as theory gone too far, but also feel that most cartographic practices, including distributed mapping, are under-theorized.

This chapter has been written to make membership of this third group seem our best choice in understanding mapping practices and their history. For we Theory Engagers, although the first two groups have produced some useful arguments, on the whole we find that Theory Avoiders are oblivious to their own theories of the correspondence theory of representation (which with Group 2 we see as discredited), while Theory Embracers too often see cartographic practices as *necessarily* technicist, militaristic, or engaging in that baleful "spatial optics" of surveillance (which with Group 1 we see as throwing the baby out with the bathwater). To engage with theory in cartography is to seek a middle ground between the non-theoretic and the overly theoretic.

Theory in the history of distributed mapping

In order to understand why we might want to be a Theory Engager in understanding the history of distributed mapping, I have employed some concepts and terminology from work by Matthew Edney (1993, 1996, 1997). Edney's work is embedded in a discourse associated in the history of cartography with Brian Harley, John Pickles, Denis Wood, the *History of Cartography* project itself (edited by Harley and Woodward), and the Monmonier of "cartocontroversies" (Monmonier, 1995), which emphasize maps as social constructions. Edney argues that the discipline of cartography has adopted a monolithic view of the history of cartographic practices. This view sees cartography as the progressive enlargement of information collected about the world – a spatial database. The database has several notable assumptions: it is scaleless; geographic facts have single geometrical locations ("location might be inaccurate or imprecise, but it is never ambiguous; each place exists in only one location" – Edney, 1993, p. 55); the data are commensurable (data can be added together or compared, and do not contradict each other – an assumption I argue leads to the current focus on "interoperability" in GIS); the database is enlarging and becoming "better" (more comprehensive, precise and accurate) over time; and the facts of the world can be read off from nature and collected (empiricism, or technically in positivism *le réel*). Note that this last assumption appeals to the correspondence theory of truth behind the map communication model and that Theory Avoiders hold most of these assumptions.

Edney argues that it is time that we drop these assumptions, because they gloss over a more productive way of seeing cartographic history as the evolution of different "modes" of mapping. Each mode of mapping is intimately tied to social, cultural and technological relations, which are contingent on particular times and places. For example, after the Renaissance the three primary modes were chorography, charting, and topography, reflecting mapping activity at various scales. By the early eighteenth century, however, these modes had merged into a single mode of mathematical cosmography (i.e., the geometrical and astronomical processes of mapping). "This merger was effectively complete by 1750: geographic data were held to be conceptually scaleless so that the scale-based distinction between chorography and special geography dissolved" (Edney, 1997, p. 43). This period of unification lasted until approximately the early nineteenth century, when cartography again fragmented into several modes, including thematic, systematic, and the revival in new form of chorographic, charting, and topographic activities.

Is progressivism in cartography simply a "straw man which can easily be knocked over" (Monmonier, 1999, p. 235)? After all, technology has yielded many benefits and advantages, including the high customization of distributed mapping, as Monmonier points out. However, non-progressivists such as Edney do not gainsay societal benefits but are concerned with the account we give of those benefits. The account they challenge says that progress takes place inevitably and linearly over time (without retrenchment, ruptures, dead ends, etc.); that it is based on a model of mapping which is empiricist; and that a database of commensurable data can gradually be built up. True, this aspect is fading thanks in part to the *History of Cartography* project, but there were many histories prior to this (and in part what it was written against – see Edney, 1999, esp. p. 2) which adopted the linear model. Some recent textbooks (e.g., Tyner, 1992, pp. 4–5) still offer a linear model of the history of cartography.

In Edney's view, *no particular mode is historically privileged over the others*; they are interrelated, contesting, and dominant at different times. Each mode may emphasize different cartographic techniques (the survey, the traverse) or different conceptions of space (geometrical, commodified, or personal). Edney's account is non-teleological in that it does not see cartography as getting better and better maps in the sense of getting our maps to reflect reality more truthfully. Instead maps are a historically contingent set of relations adapted to their environment: "a map is a representation of knowledge; the representation is constructed according to culturally defined semiotic codes" (Edney, 1996, p. 189). On this view, there is no such thing as a historically transcendent answer to the question of "what is a map". It would be impossible to give "a" definition of a map – but very easy to give multiple, competing ones (Andrews, 1996).

One reason Edney's viewpoint is useful is that it forces us to confront contemporary mapping in the same evolutionary light, and to discard determinist models of technology. Using an argument developed in a discussion of the ethics of the Internet (Crampton, 1999a) I argue that technologies such as distributed mapping should not be assigned inherent "logics" (in the case of the Internet, either that it is inherently surveillant, or inherently emancipatory) or are powers that run freely without let or hindrance, or finally that power is necessarily a negative practice. On the contrary, technologies are part of intellectual traditions, and are constituted through sets of mutual relations with society. Those relations may be constraining or emancipatory, but are not *necessarily* either. Contrary to the more provocative statements within critiques of GIS which see it as a powerfully dominating technology (Smith, 1992; Pickles, 1991) I find it more useful to think of power not as domination but as a series of

> relations (production, kinship, family, sexuality) for which they play at once a conditioning and a conditioned role . . . there are no relations of power without resistances; the latter are all the more real and effective because they are formed right at the point where relations of power are exercised.
>
> (Foucault, 1980a, p. 142)

In other words, for Foucault power does not exist without its own resistance, so that power is a negotiation between itself and a resistance. This negotiation takes place at the site where power is exercised. In the same way a technology such as distributed mapping, GIS, or the Internet is a site of negotiation and contestation between its assertion and those who resist *and* modify it, for example because it becomes an invasion of privacy. As such, it is very important to take part in this negotiation – to shape distributed mapping into the form you most prefer; what I have called being an Internet activist (Crampton, 1999a).

In examining the history of distributed mapping I therefore wish to apply the following concepts: distributed mapping is a (socially) constructed "mode of cartography" (in Edney's phrase), whose history is best written non-progressively without recourse to the empiricism of the map communication model, and without a search for "the origin" of a practice or a linear sequence of influences. Delays, discontinuities, and retrenchments are likely to be found. Power and resistance circulate through a technology and its social relations. In the next section I wish to see how these concepts play out in the particular context of the history of distributed mapping.

THE HISTORY OF DISTRIBUTED MAPPING AS A MODE OF CARTOGRAPHY

Distributed mapping is an emerging area which represents one of the most interesting outcomes of the convergence of spatial technologies such as GIS, remote sensing, and digital cartography with the World Wide

Web (MacEachren, 1998; Plewe, 1997). This convergence combines the methods and techniques of interactive mapping and spatial analysis with the distribution of functionality and resources in new and provocative ways. Two recent developments in particular have led to a surge of interest in distributed mapping by cartographers, the GIS community, and the commercial sector. The first of these is the potential afforded by "user-defined" and "on-demand" mapping functionality. User-defined mapping is the user control of data coverages, perspectives, speed of animation, etc. to be shown in the map.

Monmonier was one of the first cartographers to recognize the importance of distributed user-defined maps when he spoke of "individually tailored, one-of-a-kind maps" being sent electronically (Monmonier, 1985, p. 172). Monmonier applied the suggestions of Toffler on the "de-massifying" effects of technology. De-massification is a feature of post-Fordism (extremely flexible modes of production and labor deployment), and distributed mapping's capability to create individualized maps means that it is a form of post-Fordist cartography. This suggests interesting avenues of research into the labor practices of distributed mapping.

In a previous book (Monmonier, 1982) had also noted the outgrowth of distributed databases from remote time-sharing computing during the 1970s. In some of these early distributed databases a limited degree of interactivity and thus user-definition was possible. One of these was the US federal government's DIDS or Decision Information Display System, which had been developed by NASA, the Department of Commerce (E. K. Zimmerman), and the Census Bureau. DIDS was meant to share and distribute data to many agencies, legislators, universities, and other users (Monmonier, 1982, p. 146). Although DIDS underwent testing, due to computing hardware expense and lack of demand for its data it was never installed. Yet many of its functions can be found in today's distributed mapping systems. For example, DIDS had progressive zooms or scale changes, analogous to MapQuest's maps. DIDS was ahead of its time.

On-demand mapping refers to maps that can be created at the moment of need by the user rather than accessing an archive or map library of previously compiled maps (Crampton, 1999b). Indeed, the conditions of possibility for the creation and re-use of an archive (that is, the rules under which maps are created and archived, discussed, appropriated, forgotten, or remembered) have undergone a radical break. Maps are used quite differently in distributed mapping, as we shall see below. Transience and ephemerality mark the map: neither printed out nor saved, they exist for minutes or hours rather than centuries. Typical map libraries do not retain records of distributed maps, nor the way they are used.

In this context (the archive as the set of rules) it is useful to apply the concept of an "archeology" from Foucault (1972). An archeology is an attempt to uncover the historical rules of the formation of knowledge seen as a set of discourses. How are some things said or not said, conserved, remembered, or appropriated? Further, what are "its modes of appearance, its forms of existence and coexistence, its system of accumulation, historicity, and disappearance?" (Foucault, 1972, p. 130). Foucault's focus on discontinuities, displacements and transformations in the history of systems of thought can be utilized in the current history. In distributed mapping there is a similar epistemic break: user-defined (individualized) and on-demand (transience) mapping distinguish traditional cartography (with its emphasis on communication and static *maps*) from contemporary developments in interactive mapping and distributed GIS where the emphasis is on mapping *environments* where the maps themselves are fleeting and transient.

The second development is the capability of an interactive digital environment to handle distributed three-dimensional representations, sometimes referred to collectively as "Web3D". This latter capability has been much aided by several technical developments for world-building which can be distributed via the Internet. Three-dimensional online mapping is an extension of both traditional static 2D maps and 2D interactive online maps, whether from GIS vendors or online mapping services (Crampton, 1999c). A 3D mapping experience takes advantage of the exploratory, highly interactive nature of geographic visualization (GVis). It can also provide a "co-space" which can be occupied by more than one "avatar" or representational person, therefore allowing interaction between users. Instead of a single "best" map, a fully realized spatial environment can be created. In effect, one can enter the map itself. At the moment this remains an intriguing possibility.

In brief, distributed mapping (i.e., 2D or 3D) consists of tools, methods, and approaches to using, producing, and analyzing maps via the Internet, especially the World Wide Web. It is highly user-oriented, characterized by a distributed ability to create user-defined maps on demand. These features enable distributed mapping to be highly interactive and exploratory. Compared with traditional static maps, most distributed maps are ephemeral and transitory (e.g., neither printed nor saved) with important implications for the map archive.

DISTRIBUTED MAPPING IN HISTORICAL CONTEXT – EARLY DEVELOPMENTS

Using the notion of cartographic modes developed by Edney and discussed above, I now wish to put distributed mapping into historical

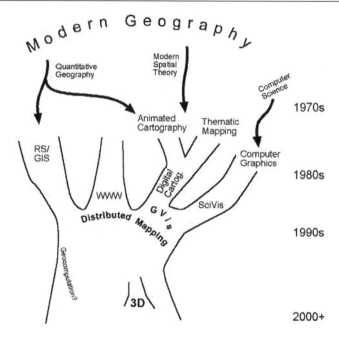

Figure 2.2   History and development of distributed mapping as a mode of cartography.

context by examining, in a partial and preliminary way, the archive. Figure 2.2 illustrates the merging and branching of the various fields which came together in the 1990s to form the current picture of distributed mapping, especially cartography, GIS, and the Internet (then later the Web). Due to space limitations my discussion will focus on only the more significant events and their implications.

CARTOGRAPHY AND GIS

Experiments in digital mapping were first made during the 1960s and 1970s. These maps were not massively distributed, but mapmaking software, for example SAS/GRAPH and SPSS, was available for mainframe computers at this time. Two events helped to stimulate this interest: the fact that geography was well into a period of intellectual growth which emphasized systematic analysis and which was later to become known as the "quantitative revolution" (Gould, 1979; Billinge, Gregory, and Martin, 1983; Livingstone, 1992, esp. ch. 9); and second, the increasing availability of computer graphic hardware (Peterson, 1995, p. 64 ff.). Geography's quantitative revolution created an intellectual space for technical enquiry, and several new graduates of departments with an emphasis on spatial analysis were important in shaping the field over the next several years. The most important of these was the University of Washington in Seattle, and later Iowa, Chicago, Northwestern, and Ohio State (which publishes the field's flagship journal, *Geographical Analysis*, 1969–). These geographers included Waldo Tobler (PhD, University of Washington, 1961), William Bunge (PhD, University of Washington, who points to Arthur Robinson for his initial thoughts on "metacartography" – Bunge, 1966, p. viii), Brian J. L. Berry (PhD, University of Washington, 1958), Richard L. Morrill (PhD, University of Washington, 1959), and William L. Garrison (PhD, Northwestern University, 1950). The work of these researchers was often informed by new modeling techniques (for example of population movements), spatial probability surfaces, and earlier works by Torsten Hagerstrand (PhD, University of Lund, 1953).

These early experiments would today not be recognized as interactive maps, and further research is needed (e.g., interviews) to determine what were the goals of these researchers. A paper by one of the most fertile quantitative geographers, Waldo Tobler (1970), provides some answers. This paper presents the results of Tobler's work on urban population growth. At the time, animated maps were investigated as ways of visualizing the solutions to geographical problems, but also as ways of exploring the data, much the same as today: "the expectation . . . is that the movie representation of the simulated population distribution in the Detroit region will provide insights, mostly of an intuitive rather than a formal nature, into the dynamics of urban growth" (Tobler, 1970, p. 238). Tobler's interest is in the geographic problem itself – in my terms, he has (3) a "mapping or visualization need" (see p. 28). Only later (late 1970s–1980s) would researchers be interested in the techniques and concepts of animated and interactive mapping per se. Tobler does not discuss interactivity, but since his movie was based on an explicit model of population, changing the terms of the model would mean changing the rate of change (for example) of the urban growth, and a new animation movie could then be produced. However, this is a far cry from interactivity as I have defined it elsewhere (Crampton, 2002a) of system response in less than a second (<1s). Tobler has, however, been credited (e.g., by Clarke, 1995, p. 5) with providing the intellectual foundation of computer cartography in his previous article "On Automation and Cartography" (Tobler, 1959) which discusses the map as part of a data processing system (the map as storage unit, as output device, etc.).

Much could be written about the quantitative revolution and its impact on cartography and computer cartography in particular. Of particular

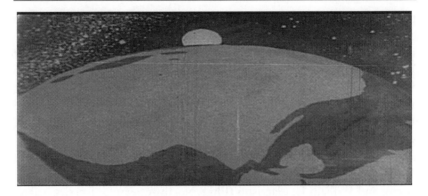

Figure 2.3   Early (1921) animated "map" featuring a spinning globe.

Source: *Animation Legend: Winsor McCay* ©2001 SlingShot Entertainment. Used with permission.

note was the early involvement of military funding agencies, such as the Office of Naval Research (ONR) which in 1959 sponsored a symposium on quantitative geography and funded Tobler and others' work (e.g., the Harvard Graphics Laboratory) through the 1960s. Tracing the influence of this funding, and development by commercial and government agencies of computer mapping software, which both stimulated and constrained developments, is a still largely incomplete task (although see the work of the "History of GIS" project, Mark et al., undated). Nor was (Anglo-American) geography alone in taking a quantitative turn during the second half of the twentieth century. A disciplinary history is in part the history of the emergence of the scientific method as the pre-eminent method of enquiry after World War II.

Looking back to even earlier efforts outside of geography, many (non-computer, non-interactive) cel animations had been produced for the public. Cel animations were first used in the film *The Sinking of the Lusitania* by the cel pioneer Winsor McCay in 1918 and shortly after in the 1921 animation *The Flying House* (a.k.a. "Rarebit Dreams"). In the latter film a flying house is shown circling a rotating earth (rotating the wrong way). It may well be the first ever animated "map" (see Figure 2.3).

In an early study Thrower (1961) examined 50 short (~3 mins) educational film sequences showing cartographic cel animations which had been made between 1936 and 1957. Although lacking many important cartographic components he did point out that animation is "unexcelled" for certain kinds of spatial relationships, especially for people conditioned on moving images in movies and TV (p. 28). However, he ended his discussion by pointing out that animation is not a substitute for static cartography, a point that could equally well be applied to today's distributed mapping. Distributed mapping should not and need not substitute for static maps.

The second development mentioned, that of sufficient computer graphic hardware, occurred in the 1960s, and marks the next step beyond frame-by-frame animation (Campbell and Egbert, 1990) into more fully featured computer cartography software. Perhaps the most influential computer program of the decade was, however, not for animation. SYMAP (SYnagraphic MAPping, i.e., "acting together graphically", Cerny, 1972, p. 167) was originally conceived in 1963 by Howard Fisher at Northwestern University and later at Harvard's Laboratory for Computer Graphics and Spatial Analysis in 1968 (Chrisman, 1988). SYMAP could perform geographic computations, such as interpolation and point to polygon (Thiessen or Voronoi) conversion, and also very valuably could use line printing to represent shading for choropleth or isarithmic maps (Monmonier, 1982, pp. 50–65). The Harvard Lab had been established by Fisher, an industrial architect, with a grant from the Ford Foundation in 1966, and was later funded by the Office of Naval Research after William W. Warntz assumed directorship in 1969 (Warntz, 1983; Mark et al., undated).

Fortunately, the history of these *technical* developments has already been quite well documented. Especially useful are Monmonier's two books (1982, 1985), Coppock and Rhind (1991), a special issue of *The American Cartographer* (Petchenik, 1988), and in the field of GIS, Foresman (1997), and the NCGIA Core Curriculum Unit on the History of GIS (Klinkenberg, 1997). Other important developments during the 1960s and 1970s include the Bureau of the Census for its DIME and TIGER databases, the CIA World Databank II – later used by the first online mapping system, the Xerox PARC MapServer, established in June 1993 – the founding of ESRI, Intergraph, and Laser-Scan. The history can be divided into several periods: the early pioneers (1960s); the role of the government agencies (1970s); and the commercial development period (1980s onwards). With the emergence of web-based GISs, we are perhaps entering a new period of "user-defined" cartography characterized by user creation of maps on demand, using highly interactive systems (Crampton, 1999b). There were many cross-linkages in cartography and GIS during this period, which indicate how technically and socially linked these developments were. For example, several researchers at the Harvard Lab moved to ESRI and were instrumental in the development of GIS there (particularly the ODYSSEY system – Chrisman, 1988).

Nevertheless, the development of interactivity in cartography did not progress smoothly from these early beginnings, and certainly did not lead neatly into the development of distributed mapping systems. In fact, the thirty years after the work of Thrower and Tobler was marked by delay and retrenchment. The concepts had appeared before inexpensive computing power was available. Certainly during the 1980s software with mapping capabilities (such as SPSS and SAS-GRAPH) was more widely distributed, but animation and interactivity lagged. By 1990, Campbell and Egbert felt so strongly about the lack of progress that they wrote a strongly critical article arguing that cartography had a long way to go if it was to do more than just "scratch the surface". This thirty years of stagnation underscores the relationship of mapping to larger societal developments (in this case sufficient computing power).

THE HISTORY OF THE WEB AND CONTEMPORARY DEVELOPMENT OF DISTRIBUTED MAPPING

The history of the Internet has received considerable attention, reflecting its high visibility among the public, journalists, and academics during the 1990s. The most incisive book on the origins and early history of the Internet is Hafner and Lyon (1996), but perhaps predictably the most detailed is an online timeline known as "Hobbes' Internet Timeline" (Zakon, 2002). I do not propose therefore to say much here about the history of the Internet, but perhaps it is worth reminding ourselves about the origins of the Web itself.

The World Wide Web (which should always be carefully distinguished from the Internet) formally originated in March 1989 in a proposal by a British physicist, Timothy Berners-Lee, working at the European Nuclear Research Center (CERN, an acronym of its name in French) in Geneva. As always, the particular circumstances surrounding it were mediated through intellectual and social connections, and its work did not progress smoothly. The original plan for the Web was an information retrieval and ordering device. During the 1980s Berners-Lee had been searching for ways of organizing information for spatially separated scientists, who used different computing environments, spoke different languages, and who worked on rapidly evolving complex systems. His solution was a distributed hypertext system that in 1989 he called "Mesh" (the term World Wide Web was substituted in 1990). Hypertext had received considerable attention in the 1950s and 1960s through the work of an independent researcher Ted Nelson, whose own work was inspired by Vannevar Bush (Director of the Office of Scientific Research and Development, and presidential science advisor) in 1945. Bush's "Memex" was not physically implemented into any working system. By the late 1980s, however, there was renewed interest in hypertext among many computer scientists, including a USENET newsgroup alt.hypertext, a special issue in 1988 of the *Communications of the ACM* (Association of Computing Machinery) and at least two conferences. Berners-Lee was aware of these developments, and modified an Apple HyperCard-like organizational system he had first developed in 1980 called "Enquire" to handle project management (Berners-Lee, 1989).

These ideas, which were put into place in 1990, did not generate much immediate interest outside CERN. After all, hypertext had been around for more than 40 years (perhaps earlier if rudimentary annotation systems such as the commentaries on the Torah are counted). What the Web needed was a way of making the ideas tangible and easy to understand. This came in the form of a graphical browser, Mosaic. Mosaic was the "killer app" for the Web, first for the X windows system under UNIX, then the Mac and Windows. It was not the first client browser (this honor again belongs to Berners-Lee, who in 1990 wrote one called "WorldWideWeb" – no spaces, later renamed Nexus to avoid confusion) but it was the first to break out to the public. Although later superseded by other products, first from Netscape, then Microsoft and AOL, Mosaic began the era of the graphical browser: the date was 1993 and the Web had arrived.

Figures for the Web are often cited in support of its rapid growth. Indeed, during the period of 1992–95, the proportion of Web traffic passing over the NSF's backbone network increased from non-existence to 26.3 percent, and in rank to first (see Table 2.1).

Table 2.1 NSFNET backbone data: proportion of traffic in bytes by port (WWW = 80, ftp = 20). Other services not listed include finger, gopher, nntp, telnet, etc.

Date	% FTP (rank*)	% WWW (rank*)
6/92	50.4 (1)	–
12/92	46.1 (1)	0.002 (186)
6/93	42.9 (1)	0.5 (21)
12/93	40.9 (1)	2.2 (11)
6/94	35.2 (1)	6.1 (7)
12/94	31.7 (1)	16.0 (2)
4/95**	21.5 (2)	26.3 (1)
[1999	~13 (n.a.)	~68 (1)]***

*   Rank of proportion of packets.
**   The NFSNET backbone was disbanded in April 1995.
*** Source: Peterson, 1999a, p. 573, percentage of all Internet traffic.

Source: Compiled by author from archives at ftp://nic.merit.edu/statistics/nsfnet/.

Table 2.2 Online population as a percentage of total world population, 1996–2005.

Date	Online popn§	% of world popn¶
1996	60	1.0
1997	100	1.7
1998	150	2.5
2000	327	5.4
2002	605	9.7
2005	720 (est.)	11.2 (est.)

§ People (in m) with access to Internet.
¶ All ages.

Source: US Census Bureau, NUA.com.

The trend has continued since 1995 as can be seen by the last line of Table 2.1. In fact, today's Internet is so congested (particularly with ".com" traffic) that a consortium of universities and business (the University Corporation for Advanced Internet Development, UCAID) has developed an advanced backbone network for "Internet2" member universities which offers sufficient commercial-free bandwidth to enable live online video–conferencing and other bandwidth-dependent scientific research. This is the Abilene project. In November 2001, Abilene achieved the first uncompressed real-time gigabit HDTV transmission across the network (Zakon, 2002).

An important proviso needs to be added, however, which is that despite this amazing growth, the Web is still only available to a tiny fraction of the world's population (see Table 2.2). This fact is sometimes forgotten in the hyperbole surrounding the Web and the Internet. Furthermore, access is highly constrained by geography, social status, age, gender, and other variables (Crampton, 1999a). For instance, the Washington, DC area has been reported as the USA's most Internet connected region, with nearly 60 percent online. Globally, the average is only 9.7 percent by late 2002.

As can be seen from Figure 2.2, the capabilities of the Internet first merge with those of GIS/cartography in the early 1990s. The first interactive mapping capabilities were established to *test* interactivity, rather than as cartographic or GIS applications as such. (We will encounter this initially practical origin again in the case of the invention of thematic mapping in Chapter 3.) It would not be until the late 1990s that distributed mapping systems were established for the express purpose of providing GIS/cartographic functionality.

The earliest map server is the Xerox PARC server developed by Steve Putz to test Common Gateway Interface (CGI) scripts via the Web, and

Map Viewer: world 0.00N 0.00E (1.0X)

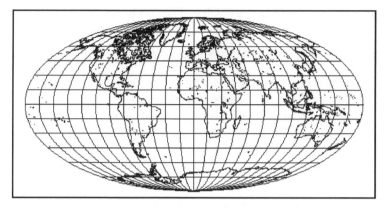

Select a point on the map to zoom in (by 2), or select an option below. Please read About the Map Viewer, FAQ and Details.

Options:

- Zoom In: (2), (5), (10), (25); Zoom Out: (1/2), (1/5), (1/10), (1/25)
- Features: Default, All; +borders, +rivers
- Display: color; Projection: elliptical, rectangular, sinusoidal; Narrow, Square
- Change Database to USA only (more detail)
- Hide Map Image, No Zoom on Select, Reset All Options

Options can also be typed in as search keywords (e.g. "lon=-100", see details). Current region is 360.00 deg. wide by 180.00 deg. (12420.00 miles) high.

Figure 2.4   Illustration of Xerox PARC map webserver, the first online mapping environment. Image courtesy of the Palo Alto Research Center (PARC).

put online in June 1993 (Putz, 1994). CGI is a method for external clients to execute commands interactively and remotely on information servers. What is interesting for our purposes is that the maps were created "on demand" by a perl scripting language according to a set of basic user inputs (latitude/longitude, scale, etc.) embedded in the URL, rather than serving images from a map archive. The on-demand maps were then served out via an HTTP process running on a Sun workstation at Xerox PARC (see Figure 2.4). Basic usage statistics indicate that it is highly popular, with some 130 m accesses since it started in June 1993 through summer 1999. The level of access has not changed substantially since spring 1996 at approximately 60,000 map images per day. For a while, the server was also accessible via a geographic name server (a service now alternatively available for US cities from the Census Bureau – see below).

Archivally, we have only a partial glimpse at the history of this innovative mapping environment. A record of the types of map in use, their geographic focus, and scale is not available. The conditions of knowing and storing this kind of information have been lost. We do not know, for example, what parts of the world are mapped or by whom. Are the maps used to explore events in the news (knowledge discovery) or to look up your hometown (knowledge confirmation)? What are the common scales used – small-scale (synoptic) or large-scale (local)?

Other significant developers were produced at the Bureau of the Census making their TIGER databases accessible online (1995) and the Digital Libraries Initiative (DLI) which aims to put a cartographic interface on georeferenced data (1995). With a DLI the user can search for images, maps, or other environmental data and metadata via a cartographic front-end. The best known is the Alexandria Digital Library (Buttenfield, 1999). Other government agencies such as the USGS also provided distributed data, though not necessarily interactively. Another well known service is "Earthview" at CERN which produces about 60,000 user-defined map views per day (Peterson, 1999a). This compares with the USGS, which prints about 500,000 non-custom maps per week and is often thought to be the world's largest print map producer.

By 1996, commercial vendors had also recognized the potential value of distributed mapping, and were offering a variety of products in the marketplace. These products fall into two categories, alluded to earlier. First, there are interactive map generators and online spatial data providers (true "online mapping"). These include MapQuest (which not only provides maps at its own site but provides those for Yahoo! and other websites as well), MapBlast! (provided by Vicinity Corp., a business services company using data from Etak), mapping services associated with online phone and people directories (such as MapsOnUs, Switchboard, and BigFoot), and a Microsoft/USGS/Compaq joint product called TerraServer. TerraServer and its ilk (e.g., SpaceImaging which runs the Ikonos program) provide detailed satellite imagery, often to about 1-meter resolution or better (Monmonier, 2002a estimates that the absolute best military imagery, the Keyhole program, is probably sub-10 centimeters in resolution). Terraserver offers a database of imagery via either a cartographic or a geographic name interface, employing declassified Russian satellite imagery (SPIN-2) for global images and USGS aerial photography (digital orthophotographs of 1-meter resolution) for the United States. According to the FAQ at its Website, TerraServer was only incidentally a spatial data provider, being developed initially by Microsoft as an experiment in terabyte (trillion byte) data scalability (a spatial database was chosen simply because it was the largest at hand).

Second, there are spatial data analysis and visualization tools available over the Internet. These offer full-blown distributed mapping capabilities rather than mapping solutions. Two developments are noteworthy: GIS companies positioning themselves to offer Web enabling of GIS, and further integration of GIS/Web/visualization technology (Cook et al., 1997) and database cross-linking (Carr et al., 1998). The latter extends the early and highly innovative work of Monmonier who first applied the concepts of geographic brushing in cartography.

Changes in the commercial sector are rapid (six-monthly cycles of change and software upgrades) and extremely competitive. I will refrain therefore from reciting specifics as this is no longer "history" but contemporary and ongoing development. It is apparent, however, that the convergence of spatial technologies is continuing.

IMPLICATIONS OF DISTRIBUTED MAPPING

Transience

One critical difference between web and print maps is their historical legacy. Web maps last for minutes rather than years. Somewhat surprisingly given their large-scale production, at any given time there probably exist in the archives far more print maps than web maps. Further research is indicated into how many print and virtual maps exist and who has access to them. However, if we distinguish between the map and a mapping environment (as I think is necessary) then it is likely that far more people potentially have access to mapping environments than print maps. It is possible to speculate that we are seeing a shift from the map (product), to the mapping environment (process).

This transience has several implications. First, the historical archive does not capture the range of contemporary mapping activities. There is a danger that many mapping practices will not be recorded in the archives. Certainly, librarians and others are keenly aware of this issue. Second, it raises the issue of what can and should be recorded. Should maps be archived, or the queries which generated them? MapQuest may generate millions of maps per day, but they are all fairly similar. Perhaps what should be recorded is the scale, region, and database query, not the map.

Cartographer/user convergence

The structure and labor processes of cartography are in transition. There is a declining need for "the cartographer" as a specialist occupation. It may never disappear entirely (should it?), but the skills and techniques of the cartographer are increasingly being migrated to the users themselves. Distributed mapping is eroding the distinction between the cartographer and the user; "now we are our own cartographers" is the oft-repeated phrase. Yet I believe that more work needs to be done in tracing out the nature and extent of this transition. Indeed, many people may want to resist it.

Map use and cognition

Distributed mapping opens up new map use environments. In recognition of this, the ICA Commission on Map Use was transformed into the Maps and the Internet commission in 1999. Of course it has always been possible to interact with maps, but now interactivity is defined as an environment where the display changes in response to user input, usually very rapidly (<1 second response time). In geographic visualization (GVis) the distinction is already made between high and low interactivity. There are many research issues which need to be pursued, including user interface studies (Torguson, 1997) and cognition. An example of the latter is the question of navigation within so-called data landscapes. Are interactive, 3D environments more efficacious in learning new environments? How does immersibility affect spatial cognition? Do map metaphors work best in organizing abstract data? One company depicts news stories as topographic maps, with local peaks (popular news stories) and valleys (less well-covered news stories).

Commercial applications

The history of GIS provides an example of how distributed mapping may diffuse away from academic geography into commercial applications. Already it is possible to see that the vast majority of web maps exist to provide support for web commerce, rather than as ends in themselves. For example, web maps are critical to the travel and tourism business, which is forecast to comprise 35 percent of Web sales by 2002.

CONCLUSION

In this chapter, I have examined in a preliminary way the history of a particular mapping practice. Taking up notions from the work of historians of cartography, especially Edney and Harley, I have argued for a type of history which is non-progressivist (emphasizing contingencies, delays, and dead ends) and which rejects the empiricist map communication model. I suggest that the history of distributed mapping marks a significant break rather than a continuation with traditional cartography. Questions of concern include what model of representation is adopted. Will it be one in which all spatial data is seen as commensurable and interoperable? Will incommensurable spatial data (local knowledges, spatial cognitions) be omitted? If so, this may simply be the continuation of empiricist progressivist cartography with its ever-increasingly accurate database under a new name.

Distributed mapping is becoming an increasingly popular mode of cartography. Much more work needs to be done to trace the contours of its development. Some immediate questions can be identified: those of the archive, labor processes and expertise, map use and cognition, and finally commercial applications and popularization.

CHAPTER 3

Why Mapping is Political

Missing from [Kant's critiques is] the mode of understanding
itself, a volume devoted to the rhetorics of how we make sense of
the world and of how we share that sense with others. Its
provisional title: *A Critique of Cartographical Reason.*

Gunnar Olsson (1998, p. 152)

HORIZONS OF POSSIBILITY

The purpose of this chapter is to examine some of the context in
which mapping is practiced and thought about. I shall make several
points. First, our present context is historical and arose from identifiable
events that help shape the way mapping takes place today. But every
context allows some possibilities and closes off others. Second, our
current context is based on a Cartesian-scientific worldview which casts
maps as communicators of spatial location. One consequence of this is
that we do not take account of maps as helping us find our meaningful
place in the world. Third, examining this context as a horizon of possibilities is itself a political project. Finally then, some possible components
of such a "politics of mapping" are sketched out that might let us understand our horizon of possibilities in order to expand it.

In Gunnar Olsson's recent essay he makes a striking observation:
modern thinking (like Kant, he calls it by the more technical name
reason) is cartographic and we need to examine this thinking, this rationality (see also Olsson, 1998). Following Olsson, this chapter asks the question: what are the historical conditions of possibility for thinking
cartographically? With this question we're asking what it's possible to
think and do. By emphasizing that these are *historical* conditions we are

not consigning events to the past, but acknowledging that different conditions may exist at different times.

I suggest that the work of Heidegger and Foucault can shed considerable light on the question of our current context in cyberspace for three reasons. First, these questions can be seen as philosophical in the same sense as Heidegger's ontological project about "being". Heidegger's constant concern with being (often capitalized in English as Being to distinguish it from *a* being) was a question not just of what exists but with being as such.[1] Being is what it means to be. Heidegger's work is notoriously difficult and strewn with vocabularies and etymologies of his own devising. Still, his ontological question does point the way towards an important emphasis on understanding being within a particular historical framework.

An ontology in a historical framework is called, reasonably enough, a historical ontology (Elden, 2001a; Hacking, 2002). A historical ontology examines the very conditions of possibility for thinking itself, in order to widen those conditions and increase the possibilities for human freedom. Putting it like this should tip us off to the fact that rather than armchair philosophizing this project is a politics, in this case a politics of mapping. Why? Because politics looks for the capability and grounds for intervening in the production of (spatial) knowledge, as well as for resistance to established power relations. Politics and philosophy (in this case ontological thinking) are both involved. Foucault's "problematizations" are another way of doing historical ontology.

Second, Heidegger's project is relevant because he argued that enquiry in general was dominated by a scientific approach which obscured essential aspects of how things are. Although Heidegger was writing in the 1920s and 1930s there is no particular reason to suspect that scientific mentalities have become less dominant, either generally or in cartography. Yet Heidegger was not anti-science. He would grant that there are many wonderful insights and achievements in science. However, these are largely confined to the physical sciences. He was doubtful whether the "human sciences" could be conceived in the same manner. Because cartography, mapping, and GIS are at the intersection of science and human science, and are also practices and technologies, it is a particularly fascinating question to see how they have proceeded in this light. Brian Harley asked us much the same question: "are [cartographers] concerned at all with how maps could answer the Socratic question 'How should one live?'" (Harley, 1990, p. 16). Or does that political question pass us by when we concern ourselves with accuracy and interoperability? Harley's own reply is pessimistic, but perhaps he was too quick to judge. In any event, the success, goals, and problems of cartography's Cartesian tradition can be assessed by Heidegger's critique.

Third, Heidegger and Foucault were acutely aware of the importance and centrality of space in their thinking. Heidegger, for example, understood our being as one of being-in-the-world and was interested in place, distance, nearness, and spatiality (see, for example, section 22 in Heidegger, 1962). Foucault's concern with space in terms of its power-knowledge relations has also long attracted interest from geography (e.g., Driver, 1985; Philo, 1992) and cartography (Harley, 1989). Both Heidegger and Foucault play key roles in a larger project of "the politics of space" (Elden, 2001a; Hannah, 2000; Harvey, 2001).

No doubt Heidegger and Foucault are not the only writers we could turn to if we want to understand mapping. What is important here are not the exact details of their writings but how their writings can cast light on mapping. The three reasons given above – highlighting the political conditions of possibility, a critique of cartography's Cartesian tradition, and a concern with spatiality – let us see what Olsson meant by a cartographic reason or rationality – our current context.

THEORY AND PRACTICE IN CARTOGRAPHY

What *is* our current context in which mapping takes place? One aspect can be examined by understanding the relationship between theory and practice. More than a dozen years ago Harley argued that cartography artificially divided theory and practice. At that time his concern was cartography's social relevance and its "theoretical isolationism" as he called it (Harley, 1990, p. 1). Mapping is often granted conceptual (theoretic-philosophic) and practical status (its practices). We distinguish between *understanding* maps and *using* maps. But there is all too often a failure to grasp how theory and practice affect each other. For example, maps are often used unreflectively for instrumental ends, to make things happen, while on the other hand some social theorists think of maps as repressive, and dangerously powerful, implying that we should use them only very reflectively.[2] In the last chapter I characterized these positions as "theory avoiding" and "theory embracing". Both have proved useful to their adherents, but both are only part of the story we can tell about mapping. Perhaps the way we use maps affects how we understand them? Perhaps then if we can't put maps into practice we gain only a limited understanding of them. (This has immediate consequences for the history of cartography because we cannot use historical maps in the context in which they originally existed.) By questioning our boundaries of thinking Harley was initiating what we can call a politics of mapping.

It's not very usual to think of mapping as a politics. That maps sometimes have a political dimension, such as propaganda maps, advocacy maps, or public participation GIS, yes. But that the practice of mapping itself (as the production of geographic knowledges) is a political project? That's not so clear. Perhaps our responsibility should be to make maps as a-political as possible. Certainly it was not too long ago when cartographers could explicitly state that there should be as little "intrusion" of politics (ideology) into mapping as possible.[3] And attention to mapping from those interested in the politics of space has also been intermittent.

The opposition and neglect of this topic arises in part from an attempt to conceptualize cartography as purely technical, but it goes further than that. It also depends on the constitution of cartographic knowledge as an a priori, that is, as beyond the reach of human conceptualizing (it existed "prior" to our concepts and politics and is independently true). With this view, maps are useful to the extent that they represent the things in the environment themselves, to cut nature at its joints (see Andrews' critical introduction in Harley, 2001). A historical ontology on the other hand suggests that the way things are, their being, is in fact a historical product operating within a certain *horizon of possibilities*. We *are* in a certain contingent way and can *be* different. If this view is valid then a politics of mapping is not just a question of propaganda maps (maps used politically) or even a political critique of existing maps, but a more sweeping project of examining and breaking through the boundaries on how maps are, and our projects and practices with them. This is politics in a very positive sense. And it would pretty much have to be a project that was always ongoing – we would never reach a conceptualization of maps "out" of history. Heidegger signals this in the title of his best-known work *Being and Time* (Heidegger, 1962).

True, beginning with the work in the 1980s of Harley, Wood, Pickles, and several others, cartography did see a burst of activity which for the first time was theoretical reflections on the field. Harley, especially, began a productive relationship with the work of several writers and texts in several fields, notably Foucault and Derrida, and Barthes, as well as work in semiotics and iconography. But Harley's project was sadly incomplete when he died in 1991 (at the age of just 59 years), and perhaps more importantly, mistakenly founded as the recovery from "behind the text" of the map's silences and secrecies (Harley, 1988b). He was not therefore able to draw together these disparate theoretical explorations on the (political) power of maps. Or rather, before he was able to conceive of the power of maps as a political project. Harley's great contribution was to show how maps have been used politically, as political documents, and that such use of maps is of valid concern. This contribution can be summarized as how the political has entered and been deployed in mapping.

An illustrative example of the political for Harley lies in his interest in the Peters projection. Harley strongly resisted the then prevailing attitude that the Peters projection should be criticized on the grounds that it had political purposes (a rejection I've argued was often cryptically hidden beneath criticisms of the map's aesthetics or origins – see Crampton, 1994). Indeed, Harley was delighted how the projection had ruffled feathers in the cartographic literature. But even in Harley's work there is a strong conception of maps as documents which provide ("confess") the truth of the landscape, a truth to be sure often repressed by political intervention, but nevertheless there to be recovered from its "subliminal geometries". Therefore we can trace his project to exactly the same one which founds cartography as a whole, and which is captured in Monmonier's axiom mentioned at the outset of this book: "not only is it easy to lie with maps, it's essential" (Monmonier, 1996, p. 1).

While Harley (and others in his wake) provided an examination of the political in mapping, that is to say, how maps are employed as political documents, this totally evades the question of how mapping *necessarily produces* the political, and how rethinking mapping can lead to a rethinking and questioning of the political. On the Harleian agenda the political is assumed unproblematically and the history of mapping is privileged from *within* a discipline rather than questioning ("problematizing") the condition of possibility for thinking, a spatial politics.[4] Foucault's conception of his project as a problematics arises from Heidegger's distinction between *ontic* knowledge of things in themselves and *ontological* knowledge, or the conditions of possibility for ontic knowledges. Our day-to-day understandings of the world (ontic knowledge) remain unquestioned and unthematized until a crisis or breakdown (a problematics) precipitates an examination of the domain of enquiry. As Schwartz (1998) aptly puts it: "the notion of problematization can be grasped as a creative reworking of Heidegger's account of equipmental deficiency" which means that "we cannot even *think* about our ways of existence until they have become problematized, much as Heidegger posits that *Dasein* becomes conscious of objects only in the advent of an equipmental breakdown" (p. 19, original emphasis). This *ontological* examination is necessarily philosophical, because for Heidegger just such a crisis has been reached with the human relationship to being; and since all such breakdowns are reflections of this founding relationship it becomes the business of philosophy. As long as a domain of enquiry, such as cartography, is concerned with how it should proceed from within its own framework (what color schemes to use on maps, the effects of scale, how maps are used historically or politically) then it is engaged ontically. It is only engaged ontologically when it faces the historically contingent conditions of possibility (the horizon of thought) and it is political insofar as it discloses and challenges those contingent limits. This as such and of itself is both a definition and a call to a project of a critical politics of cartography.

Harley's attempt to address these issues (e.g., Harley 1988a, 1988b, 1989, 2001) was a necessary step in bridging the intellectual gap between theory and practice. Harley was a new kind of cartographer, and I can think of few other cartographers before him who studied the relation between maps and power. Symptomatically he would deny he *was* a cartographer, but in fact his work can also be understood as making it possible to *be* a new kind of cartographer. But even Harley constituted cartographic knowledge as a priori. Thus his project became one of uncovering the layers of ideology inscribed in the map to get at the golden nugget of truth underlying it all. On this view, power is repressive (a view never held by Foucault). Since Harley's death, progress toward a critical politics of cartography which bridges the gap between theory and technology has been sporadic or carried out under other names – yet it has never entirely disappeared (see Yapa, 1991, 1992; Edney, 1993; Pickles, 1995; Cosgrove, 1999; Harvey, 2001; Black, 1997; Elden, 2001a; Monmonier, 2001). What is at issue, but which has not yet been clearly articulated in this work, is a critical politics of mapping, rather than just a political critique of existing maps (more on this below).

Harley's fruitful contribution was to ask the vital question about what mapping is and could be, and like Heidegger to set us on the path of questioning its possibilities. That's why Matthew Edney called his obituary of Harley "Questioning Maps, Questioning Cartography, Questioning Cartographers" (Edney, 1992). But surely other cartographers and geographers have also thought about what mapping is? Arthur Robinson, for example, even co-authored a book called *The Nature of Maps* (Robinson and Petchenik, 1976). Was this not about the being of mapping? The longer answer to this is suggested in the next section but the short answer is that Robinson and Harley's projects were different because Robinson tried to describe how maps are, whereas Harley describes why maps are as they are, and how else they can be. It is only this latter project which is the political one.

It is a key argument of this chapter that maps and GIS are important sources for the production of geographic knowledge. What are the power-knowledge relations of mapping as they occur against the historical horizon of possibilities and how that horizon can be enlarged? This is a question of the historical formation of mapping concepts (e.g., about cyberspace) as an epistemology, and the possibilities that are given to us for the being of those concepts, or an ontology. In other words, theory and practice.

"THE FISHERMAN'S PROBLEM": ONTIC AND ONTOLOGICAL KNOWLEDGES

What does it mean to open the question of the conditions of possibility for cartography, and how does this constitute a question which is philosophical and political? To provide an initial response to these questions we can go back to the Heideggerian distinction between the two types of knowledge (see Heidegger, 1962, §4):

1. ontic knowledge, which concerns the knowledge of things as such; and
2. ontological knowledge, which concerns the conditions of possibility for ontic knowledge.

For example, the question "how old is the Vinland map?" is an ontical question, whereas "what is the mode of being of maps?" is an ontological question. The first question may be addressed and resolved by science, but not the second (Polk, 1999, p. 34). Elden elaborates that "Heidegger's own exercise of fundamental ontology deals with the conditions of possibility not just of the ontic sciences, but also of the ontologies that precede and found them. This is the question of being" (Elden, 2001a, p. 9). Heidegger's distinction suggests that ontical enquiry often characterizes disciplinary work because it can be addressed scientifically. In the discipline of cartography, for example, we might enquire how to satisfactorily generalize and symbolize landscape features, or which projection best reduces distortion. But this ontic language of science and objectivity itself takes place within a conceptual framework (ontologically). We can call this "the fisherman's problem", using an insightful metaphor from Gunnar Olsson: "The fisherman's catch furnishes more information about the meshes of his net than about the swarming reality that dwells beneath the surface" (Olsson, 2002, p. 255). The fisherman certainly catches real fish that were in the ocean (that is, ontical enquiry certainly can say truthful things about the real world). But if he tried to say something about the reality of the denizens of the ocean, his explanation would be related to the size of his fishing net. He wouldn't have much to say about whales or sharks, nor about sea anemones. The net therefore plays a double function of both revealing things about the sea and hiding or concealing them. For Heidegger this double function of unconcealing–concealing is an abiding aspect of our understanding of being. If Heidegger is right then studying maps and mapping would seem to include as much about what maps can't or don't do as what they can do. This is why Harley spoke of the silences of the map (Harley, 1988b).

If we now go back to the difference between Robinson and Harley we can see that where the former described the fish in the net, the philosophies of Foucault and Heidegger are concerned with the net itself. Harley also asked about the net. What does the net catch? Do we like what it catches? Have other places or times had other kinds of nets which caught different things? What do we suspect the net to be unable to catch? How can we change the net to catch other things? According to Heidegger our present "ontological net" is critically flawed because it sets up being in a very scientific way. We like to measure things and treat them as objective presences on the landscape which can be re-presented. For Heideggerians, this understanding of being constitutes the unfortunate "metaphysics of presence", a kind of planing down of the full measure of being. In Heideggerian terms "we are neither present *substances*, nor present *objects*, nor present *subjects*" (Polt, 1999, p. 5, emphasis added). This critique of mapping as a rigid staring at something objectively present (cf. Heidegger, 1962, p. 61) should remind us more of Harley than Robinson.

The ontic–ontological distinction is a familiar one in the history of philosophy, dating back to Descartes and Kant. When Heidegger took it up, he distinguished between living life as such (making choices against a background of possibilities) for which he coins the term "existentiell" understanding, and the questioning of what constitutes existence and the structure of these possibilities, which he calls the "existential" understanding (Heidegger, 1962, §3–4). This *existential* understanding is one directed toward the meaning of being. Heidegger begins his book by stating that we are very far from answering the question of what an existential understanding might be; so far, in fact, that the very question itself is forgotten (§1).

These bewildering terms might make us wonder why it's worth worrying about the "being of maps". Why not study concrete maps that actually exist? Heidegger's response is essentially to refer us once again to the fisherman's problem. Sure, we could study the contents of the net. This is what we do when we study maps and mapping, especially from a scientific viewpoint. It is ontical enquiry about things. But the only way to know anything meaningful about the nature of the ocean is to understand our conceptual framework from within which we understand that ocean – to look at the net itself. This ontological looking means thinking about being as such, including the being of maps. The fact that it sounds strange to say this ("the being of maps") is just one indication that we hardly ever think this way, that is, philosophically. Perhaps if we do so, we can open up a new and productive dialog about mapping.

HOW WE MIGHT DO PHILOSOPHICAL THINKING

> What is philosophy today – philosophical activity I mean – if it is
> not the critical work that thought brings to bear on itself? In what
> does it consist, if not in the endeavour to know how and to what
> extent it might be possible to think differently, instead of
> legitimating what is already known?
>
> Foucault (1985, pp. 8–9)

If we grant that the ontic–ontological distinction is helpful, it is still not immediately apparent how ontology might be carried out in cartography. And what about ontical enquiry? If the whole way maps can be is expanded, it seems as if the ontical questions would have to change too. Following from the earlier discussions which initially introduced problematization, we can now sharpen and fine-tune these ideas for mapping cyberspace. Since ontological thinking is rare and neglected (according to Heidegger) there won't be many examples to draw from. Luckily there is one well-known example that we can examine which picks up where Heidegger left off. Even better, it is directly relevant to cartography.

The following is an extract from a lecture in 1983:

> Most of the time a historian of ideas tries to determine when a
> specific concept appears, and this moment is often identified by
> the appearance of a new word. But what I am attempting to do as
> a historian of thought is something different. I am trying to
> analyze the way institutions, practices, habits, and behavior
> become a problem for people who behave in specific sorts of ways,
> who have certain types of habits, who engage in certain kinds of
> practices, and who put to work specific kinds of institutions. The
> history of ideas involves the analysis of a notion from its birth,
> through its development, and in the setting of other ideas which
> constitute its context. The history of thought is the analysis of the
> way an unproblematic field of experience, or a set of practices
> which were accepted without question, which were familiar and
> out of discussion, becomes a problem, raises discussion and
> debate, incites new reactions, and induces a crisis in the
> previously silent behavior, habits, practices, and institutions. The
> history of thought, understood in this way, is the history of the
> way people begin to take care of something, of the way they
> became anxious about this or that for example, about madness,
> about crime, about sex, about themselves, or about truth.
>
> (Foucault, 2001, p. 74)

It is worth trying to understand Foucault's meaning here. He begins by making a claim that the work he is doing is a history, but that it is not like the history we are most typically used to. So Foucault is a historian but not a traditional one. A traditional historian is interested in the "history of ideas" or what is thought at a particular time (the zeitgeist, contemporary discourse, what people said at the time as recorded in newspapers, journals, writings, records; i.e., the historical "archive"). Foucault, however, is interested in how things "become a problem" or problematizations. When something which was previously unproblematic does become a problem then people start to pay attention to it, even worry about it and try to deal with it. We can pick up on these periods of problematization as times when the regular ongoing behaviors are no longer possible in the old way. It might cause "cartographic anxiety" as Gregory called it (Gregory, 1994). In this sense, mapping is a problematization itself. We map because we are concerned with a certain aspect of the environment and wish to try and deal with it. A Foucauldian history of cartography would be a history of how a particular problem was taken up cartographically.

In fact, it's the fisherman's problem again. We reel in the net and find it has big gaping vents and weird bite marks over it which prevents us from fishing as normal. We begin to suspect some large beast down there that is too strong for the net, so we research ways of strengthening the net or making the mesh more coarse. Or perhaps we switch from net fishing which scoops up everything to making a distinction between fish-for-consumption and fish-as-part-of-an-ecological-system. Now fishing is not just a question of extracting resources but concernful participation in an ecological system. Because of a problematization, fishing as a way of being has changed.

A good cartographic example is provided by the controversy over the Peters projection. In the decades following the introduction of his world map, Arno Peters attracted dozens of articles which were highly critical of it (Monmonier, 1995). But Peters persisted, his map was adopted by aid agencies and the World Council of Churches, and was featured on the US TV show *The West Wing*. It was and is still a big problem for cartographers. While their approach was ontical (they pointed out all the technical reasons he was wrong) it is also possible to read the controversy as saying something defining about cartography itself (Crampton, 1994). Perhaps Peters, explicitly using the map as a politics, has made a new way for mapping to be. On this view, the cartographic opposition is inadequate, not because cartographers missed the point (their technical criticisms of Peters were certainly true) but because Peters created a new point! Peters had created a new ground for mapping – an ontological achievement.

We saw in Chapter 1 that problematization is to first set something out as an issue, second to undertake a history of thought rather than ideas, and third to examine the larger truth claims of discourses. We can now see therefore that problematization is an analysis of the conditions of possibility for ontic knowledge. It is an ontological ground-clearing. It is also necessarily political in the sense offered earlier – the conditions and possibilities of being in place. Often these conditions remain unanalyzed and only at certain times do we question our horizons of thought.[5] This has many fruitful ramifications, not all of which can be examined here. One important aspect, however, is that every context establishes normalized ways of being. The hue and cry over the Peters projection for example was over whether it was acceptable for a map to be like that. Normalization is a very powerful aspect of ontology because it tends to stabilize established power-knowledge structures. Normalization is often one of those negative effects of power with which Foucault is identified. When people especially are on the wrong end of normalization processes it can ruin their lives, but the response to this is not to escape from power but rather to use it productively (McWhorter, 1999). Power's positivity is an aspect of Foucault's enquiry often overlooked.

PROBLEMATIZING THE ESSENTIAL LIE

In this section I would like to contrast and play off against each other two books by Monmonier (1996, 2001). In the first we can analyze his assertion concerning the lie of the map to show that this very powerful statement pervades cartography, and that it produces the unproblematized ontology of contemporary mapping. By contrast, in Monmonier's more recent book (Monmonier, 2001) it is possible to discern some pointers towards a more critical problematization of cartographic knowledge production.

Monmonier (1996) writes "[n]ot only is it easy to lie with maps, it's essential" (p. 1). There are at least three terms of significance: "easy", "lie", and "essential". All three terms surround a fourth, the map, which takes its shape and its being from this tripartite structure in which it finds itself. It is of the essence, it is essential, necessary, that maps lie. In order for a map "to be" a map, it must lie. Lying is in the essence of the map. Furthermore it is easy for maps to lie, it is not something which is difficult or which can only be achieved after a struggle in the sense of going against something's nature. This ease is well-known and assumed in the statement which could thus be rewritten: "Not only (as you know) is it easy to lie . . "*but also* (and here we introduce the new idea, which we didn't previously know) it is essential and necessary. The natural ease of lying becomes something that is essential and important, that is we don't have to struggle against this natural tendency of lying, but rather should embrace it as something positive. This is further alluded to in the next few lines where Monmonier writes that "to avoid hiding critical information in a fog of detail", in order that the truth does not get overwhelmed "an accurate map must tell white lies" (p. 1). So this positivity, this advantage to lying, is that it will yield truth. In order to tell the truth, we must lie. So any truth-telling, such as the map, comprises as an essential part, lie. A map is both lie, and necessarily and as a result, truthful. And "there is no escape" (p. 1) from this.

This is an old and essential idea in cartography. It can be found, for example, in the famous saying of Korzybski that "a map *is not* the territory it represents, but if correct, it has a similar structure to the territory, which accounts for its usefulness" (Korzybski, 1948, p. 58). The political consequences of this ontology of mapping are clear.[6] Our task as mappers becomes one of deciding where to draw the line between the elements of truth and lie in the map. It is a normative ontological statement: maps "ought to be" truth-tellers. We police the boundary, we watch it, in order to make sure that there is not too much lie nor insufficient truth in the map. It becomes a question of separating the good maps, where the lie can be justified (it is just, legal) from the bad maps, where the lie cannot be justified (it is illegal, it has passed over the horizon). In this way we make the difference between USGS topographic quadsheets and propaganda maps. We're immediately made aware of the danger of sliding away from truth-telling by Monmonier: "it's not difficult for maps also to tell more serious lies" (p. 1). Thus in order to recognize when a map moves illegally across this border Monmonier has written this book, a text on drawing the line which is therefore an ethical text on the problem of truth in mapping. These "dividing practices" of normalization are a hallmark of modern thought. Dividing practices are normal and indeed *normalizing* in cartographic thought.

Monmonier's fascinating account of the role of mapping in producing favorable electoral districts (Monmonier, 2001) illustrates the difference between acknowledging the use of maps for political purposes and a more strongly conceived critical politics of spatial knowledge production by mapping. For most of his book, Monmonier discusses redistricting, gerrymandering, and the legal requirement that political districts achieve "compactness" (as measured through competing indices), and how these distinctions can be used to derive legitimate from illegitimate voting districts (in the legal sense). As such, his discussion is an example of how ideas act as dividing practices, especially between what is acceptable and what is not. However, in the last two chapters Monmonier turns from this historical account to explicitly question the way that modern voting districts are constituted.

Using the nomination (later withdrawn by President Clinton) of Lani Guinier to assistant attorney general for civil rights in 1993, Monmonier points out that alternative methods of electing representatives – multimember districts and proportional representation – to the American (and the UK and French) system of "first past the post" have plenty of historical and international precedence. According to Monmonier "[p]roportional voting is used extensively throughout the world, by developed countries in northern Europe and the western Pacific as well as by less prosperous nations in Latin America and parts of Africa" (2001, p. 144). Thus, despite the negative press Guinier received (as a "quota queen", and a promoter of racial preferences) Monmonier interprets her as problematizing the political agenda as far as space and representational politics are concerned: "American-style elections are not a prerequisite for democracy" (p. 146). This raises the question of what prerequisites are necessary, and what the historical horizon of possibilities might permit or disallow at the moment.

Monmonier successfully "puts into play" questions concerning space and politics in real-life practical situations. As such, his work is potentially useful for a critical politics of representation and mapping, and for critical geography more generally. Monmonier does not necessarily cast his work in this light himself. But thinking critically and philosophically about mapping, space, and politics does not necessarily entail taking up a position on the political spectrum. It is rather to question the essence of that spectrum and to help redefine it.

TOWARDS A CRITICAL POLITICS OF CARTOGRAPHY

> Even . . . apparently arcane ontological and epistemological
> questions must be part of the debate [about cartography]. They
> too raise issues of practical ethical concern. Our philosophy – our
> understanding of the nature of maps – is not merely a part of
> some abstract intellectual analysis but ultimately a major strand in
> the web of social relations by which cartographers project their
> values into the world.
>
> Harley (1991, p. 13)

Harley's words suggest that it is but a short step from questioning the bounds and limits of our lives (philosophy) to politics and that maps are an important practical component of social relations. It is an important step which connects philosophy and action. Why is this?

First, maps might be sites for and outcomes of struggle. This struggle is a political one where knowledge and power structures meet. To understand cartography politically opens and allows intervention in the struggle over the deployment of power-knowledge effects. On the basis of these questions it is possible to imagine *new* possibilities, changes, and human being at both the individual and societal levels for cyberspace, as Guinier and Monmonier indicated. As such, this is a political project where we see "the development of domains, acts, practices, and thoughts that seem . . . to pose problems for politics" (Foucault, 1997a, p. 114).

Second, maps may not have to produce space only objectively and scientifically, as a Cartesian set of things located in space, which has nothing to do with how we live, our experiences, or pleasures. Maps ought to be able to play a significant role in the political project of finding our place in the world. As Harley put it "when we make a map it is not only a metonymic substitution but also an ethical statement about the world . . . [it] *is* a political issue" (Harley, 1990, p. 6, emphasis added). For example, if we are interested in understanding a historical map we may think we need to examine it as an object and to assess what information it may contain (e.g., see Woodward, 1974). Yet this will not tell us how the map was used and lived as part of a struggle of making sense of the world. It will omit the experiential side of the map as well as any lived context in which to situate our understanding. Using or experiencing historical maps in their original context is not easy. It's no wonder that instead we objectify maps. Yet maps are meaningful understandings of the world, not just mechanisms for communication. This point echoes a critique made as long ago as 1976 by Leonard Guelke (Guelke, 1976). Guelke argued that the focus on communication in cartography was seriously inadequate because it doesn't take into account map meaning.

Insofar as a map is thought of as simply communicating an already known and digested *knowing*, then the questioning (of the horizon) is not permitted and is foreclosed. This very foreclosing gives the map its authority and power. But "it awakens nothing in the way of a questioning attitude or even a questioning disposition. For this consists in a *willing*-to-know. Willing – this is not just wishing and trying. Whosoever wishes to know also seems to question; but he does not get beyond saying the question, he stops short precisely where the question begins. Questioning is willing-to-know" (Heidegger, 2000, p. 22). If we use a map just because we wish to know something, to be on the receiving end of an information transmission, then we have stopped short of mapping as problematization. We have chosen to limit ourselves to thinking within the bounds of our ontology, rather than willing to know what mapping can be and how it can open up a world. The map has become an object in our world rather than a political site.

In the ontic cartographic practice so far established the best maps are those which are the most conclusive, the ones which most authoritatively communicate the truth of the landscape (an authority which is vested in their adherence to the rules, rules which are at this particular historical juncture provided by science). But what we aim for here are maps which willfully challenge normalization. For from this questioning comes the possibility of an unfolding of the being of maps and mapping. In the remainder of this chapter therefore I wish to suggest or open up some possibilities which might contribute to a critical politics of cartography by posing two major questions: why pursue a critical politics of cartography; and second, of what does it consist?

Why pursue a critical politics of cartography?

We can begin this question by identifying a necessary linkage between the political and the spatial, a linkage which is essential, rather than just an occasional political option. The manifold relationship between space and politics has been examined elsewhere (see, for example, Elden, 2000) but we can gain a flavor of it by returning to the origin of the word "political". What did this word mean for the Greeks? As Sallis puts it, referring to Plato's cosmological dialogue the *Timaeus*:

> How is it, in particular, that reference to the earth belongs to political discourse? The answer, most succinctly, is: *necessarily* – taking necessity to have the sense it has in the *Timaeus*. Discourse on the city [*polis*] will at some point or other be compelled, of necessity, to make reference to the earth; at some point or other it will have to tell of the place on earth where the city is – or is to be – established and to tell how the constitution (*politeia*) of the city both determines and is determined by this location.
>
> (Sallis, 1999, p. 139)

The political then originally meant how we should live, and how we should arrange the city (or place or site) in which we need to live. To use Heidegger's phrase, we are concerned with our being-in-the-world. At the beginning of this chapter I suggested that Heidegger brought a geographic sensibility to light, and here we can see why. The spatial in the sense of this *polis* constitutes the political. Here we are very close to the phenomenological tradition in geography. The phenomenological tradition in geography is now several decades into its project, although the variety and richness of this writing is by no means all reducible to Heidegger (e.g., see Christensen, 1982; Pickles, 1985; Schatzki, 1991;

Harvey, 1996, esp. pp. 313ff.; Peet, 1998, esp. ch. 2). Phenomenology is also informing some very interesting reconceptions of the politics of space and cyberspace (e.g., see Casey, 1997; Coyne, 1998; Elden, 2000, 2001a; Sallis, 1999). Some of this work has dealt with "khora", the Greek word traditionally but problematically translated as "space" (see Derrida, 1995; Grosz, 1995; Sallis, 1999; Heidegger, 2000, esp. pp. 69–70). Other works have taken up Heidegger's challenge to rethink the *polis*, or as he would have it, the *site* of our being and thus the political (Heidegger, 2000, p. 162). For example, Elden (2000) argues that because Heidegger explicitly links the political to the geographic there is "the potential for rethinking the essence of the phrase 'political geography'" (p. 409). These thoughts have some bearing on authenticity and the politics of space in cyberspace.

Elden claims that:

> In his rethinking of the πόλις [*polis*], Heidegger makes a
> potentially major contribution to political theory, by suggesting
> the links implicit in the phrase "political geography" . . . following
> Heidegger, we might suggest that "there is a politics of space
> because *politics is spatial*".
>
> (2000, p. 419, original emphasis)

Elden's work (see especially Elden, 2001a) is critically important here because he recovers from Heidegger the idea of the *polis* as the site of human existence (an idea which was lost when *polis* was simply translated as "city" or "city-state"). The *polis* rather is the site and abode of human history. It's the "world" of being-in-the-world. As a spatialized entity (site, abode) it is what constitutes the political and allows us to rethink it. Maps, because they "make reference to the earth", are part of this constituting. Maps *produce* knowledge through mapping practices, but as problematizations their knowledge is always in a certain context, is normalized, in a power relation, and therefore for all these reasons, political.

In other words there is a necessary relation between the site of human being (the *polis*) and its spatially grounded condition. For Sallis the *polis* dialectically produces place as much as place produces the political, but it is only this latter sense which we find in Elden (see, for example, Elden, 2000). Both Elden and Sallis make a necessary relationship between space and politics, but while Sallis translates *polis* as city, Elden warns this should not be interpreted through the modern conception of the city or state: "the modern concept of the state is . . . remote from the Greek *polis* – site" (2001b, p. 325). Instead of using modern politics to understand the *polis*, we should use an understanding of the *polis* to rethink politics.

Figure 3.1 In this Doonesbury cartoon, the joke is dependent on a distinction between the content of the map being political (caribou-as-Democrats) and the position of the map within a political situation and how it helps constitute that political situation. *Doonesbury* © 2001 G. B. Trudeau. Reprinted with permission of Universal Press Syndicate. All rights reserved.

It is not a question therefore of examining "the" political in mapping, which is how the question has been framed until now. It is not a question of "looking for" the political in maps, for this would be to assume an a priori realm of the political which is sometimes injected into maps and which makes their content political. On this view we are mislead into uncovering this political content, which is the project I argue Harley pursued. On the view I am discussing here, the project is rather to investigate and reveal how mapping necessarily produces the political, and how rethinking mapping can lead to a rethinking and questioning of the political. This as such and of itself is both a definition and a call to a critical politics of cartography.

A Doonesbury cartoon can bring to light some of these points (see Figure 3.1). In 2001, a USGS cartographer lost his job over a map he made of caribou calving areas in an area wanted for oil exploration. Rick says "it [the map] was political". The joke is that Joanie deliberately misunderstands and thinks that the *content* of the map (the caribou) is political (obviously caribou can't be Democrats or even Independents). What *is* political is the map's position in a wider political situation. This example shows that a politics of cartography does not study the political content of the map – as if we could temporarily "adopt" a political mode of enquiry or "look for" political things in the map (as has happened in studies of propaganda maps) but how maps as spatial knowledge creatively constitute politics itself. Our target is politics (understood as a horizon of possibilities) and not maps themselves. This is why it is "a politics of mapping" and not a cartography of politics.

In the last section of this chapter I will sketch out a few possibilities for what a critical politics of mapping may look like. These are not propositions, axioms, or even guidelines, but rather some issues that might bear thinking through. The idea here is not to put boundaries on a subject, but to open up and explore it. Perhaps they are best seen as statements in the process of being superseded, overturned, and rejected.

Of what would such a project consist?

A critical politics of cartography is a problematization

As we have already noted and led up to, a critical politics of cartography is highly situated spatially. That is, specific understandings of space at particular historical moments are analyzed. A problematization of these moments would enquire what issues were taken up as problems in order to investigate the horizons of possibility of mapping. For example, why did thematic mapping emerge in the late eighteenth and early nineteenth centuries in Europe (especially France)? During this period (1780–1830) many of the standard thematic map types we are familiar with today were invented, such as the choropleth and proportional symbol maps. An enquiry about the conditions under which these map types were invented might proceed from the fact that they were not invented by cartographers but were part of a specific discourse about political economy. Thematic maps were instrumental in forming a statistical framework in which to understand the problem of governance. Statistics were increasingly used to assess "moral" questions, or what we would now call socio-economic issues (crime, birth rates, suicide, early marriages, etc.).

Statistics were able to provide insight into what was "normal" and what was abnormal or deviant, and maps were then able to produce pictures or snapshots of normality over the territory of the state. This led in part to an increasing need to collect more statistics, and the nineteenth century saw a great boom in these statistical collection procedures, most notably of course the national census (Hannah, 2000, 2001). Atlases of the census, such as Francis A. Walker's great atlas of 1874 (the first statistical atlas of America) were extensions of this way of producing geographic knowledge (as normalized resources). What was a problem for the nineteenth-century political economists was the issue of how best to govern the territory of the state and it was operationalized in a very particular way which has had long-lasting effects (not the least of which is the predominance of statistical mapping in problem-solving). A critical politics takes up the way that maps have been cast in an effort to imagine other cartographies that are not based on mapping normalized resources. We saw this earlier when we encountered Heidegger's critique of science as an ontic enquiry. Problematizations are concerned with the ontological horizon of possibilities.

Critical politics of cartography is a struggle in the sense of a political intervention or participation

A critical politics is not passive, but also very actively directed at intervening in the production of cartographic knowledge. This arises because as a problematization we are interested in how the particular historical horizon came to define our thinking and practices. As we have seen normalization is one powerful procedure in stabilizing this horizon, a stability which can nevertheless by undermined through a critique which sees the horizon as contingent and changeable following intervention. Public Participation GIS (PPGIS) and community mapping (Aberley, 1993) have developed ways to empower communities, especially those which are marginalized.

These efforts can be large and strategic, or small, tactical interventions on a particular issue. As an example of the latter, in the spring of 2002 a small graduate cartography seminar performed community mapping in an Atlanta neighborhood called Cabbagetown (Crampton, 2002b). Cabbagetown is one of Atlanta's oldest working-class neighborhoods, founded in the 1880s as a factory village to support a cotton mill. Today its very identity is being contested as it undergoes gentrification and the conversion of the mill into gated lofts. The seminar was interested in how the historical "memory" of this unique neighborhood may be expressed through mapping, and how in turn those memories may be spread and made accessible to current residents. The solution involved an online GIS, resident surveys, participant observation, and many other ethnographic practices. The goal is to work with community leaders and residents in order to make the online GIS part of the experience of living in Cabbagetown (as opposed to an outsider's representation of it). In this sense, mapping is a struggle over how to remember the past and to write its biography in maps. Often this writing means opposing received wisdom or the "auto-bio-geographies" inscribed by structures of power. Thus in general we can say that a critical politics of cartography involves the positive production of counter-memory (McWhorter, 1997) and counter-mappings (because they are written counter to power).

The critical politics of cartography is an ethics

It is what Foucault (1985) called an *askēsis*, a Greek word for exercise or practice. That is to say the project is "ethical" if by this word we understand not the "rights and wrongs" of mapping, but *ethos*, the mores or practices of the time. Ethics means: how shall we live in practice? What is the origin of these practices? How do they constitute the horizon of possibilities of being? What other mapping practices might emerge under a different horizon and how can we open these other horizons? Mapping practices as an ethics in this sense have yet to be properly considered politically. One suggestion is to take up the challenge of the ethics of mapping as a practice of freedom (Foucault, 1997d) through the "pleasure of mapping". Given how desire has operated to so completely normalize people, for example "gay desire" (McWhorter, 1999), it may be that sheer pleasure offers some positive ways forward. Maps as pleasure is appealing, perhaps evoking the reason people take up mapping as a practice in the first place, before it is laden down with jargon. It is in this sense that I use the phrase "maps as finding our place in the world", maps as pleasurable sense-making of the world. Unfortunately we still know very little about the pleasure of mapping – although Wood has written on it (Wood, 1987) and Harley's beautiful piece on the map as biography may hold some initial clues (Harley, 1987b; see also Gould's response, 1999, pp. 74–8).

A critical politics of cartography is a technology

By this I mean that we engage with the specific technological question of cartography and its relation to power-knowledge. As was mentioned earlier, cartography raises this issue to the foreground because of its singular place at the intersection of art, science, technology, and practice. In today's context by "technology" we mean primarily cartography and mapping as ways of being that depend on instruments and digitality as a means to an end. As such, it may leave behind other aspects of "technology". Recall that the original word for technology is the Greek *technē* which meant art, skill, way of making or doing. This sense is, however, quite lost when mapping technology produces knowledge as a resource or "standing-reserve" (Heidegger, 1977). Two short examples illustrate this point.

First, the question of "interoperability" or how well data and databases integrate with each other. Interoperability has been mentioned as one of the leading technological issues in GIS and digital mapping today (Monmonier, 1999) although the word only came into common usage in the early 1990s (in the sense of integrating software or data; the word was used prior to this in a military context to refer to how well military equipment from different countries worked with each other, as well as how different computers networks can be integrated, but these are not necessarily the same associations we have in GIS/mapping now). What role does interoperability have on the normalization of data? For example, what value will be attached to data that cannot be made interoperable (because they are too local or outside the scientific purview)? How will we judge and value maps or databases when they already have an a priori

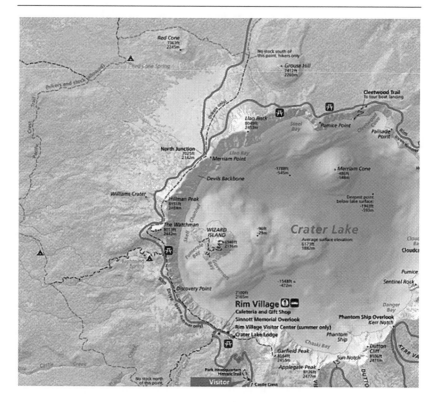

Figure 3.2   Crater Lake, Oregon on National Park Service (NPS) maps. Produced entirely from digital sources. *Source*: Patterson (2002).

existence as interoperable? This is technology as an impoverished instrumentality because it is a cause of an end already in sight (that is, interoperability). What we are interested in with technology, however, is how it can bring about insight into meaningful human life.

Second, the relationship between cartography as a science and an art is still seen as problematic (see, for example, Woodward, 2001a). No doubt this is part of a larger question of the degree to which cartography sees itself as a technology, science, or art. Some cartographers have demonstrated how it is possible to productively reinterpret technology not in order to exploit the environment as a resource but to let the essence of the landscape emerge (e.g., Patterson, 2002). Patterson has mastered the art of digital mapping from a manual tradition which pays close attention to the things themselves (see the work of Erwin Raisz and Heinrich Berann). An example of his work is shown in Figure 3.2.

Figure 3.2 shows the detail and realism on NPS maps, the aesthetic quality that can be achieved in the digital realm and not least the superb integration of art and science by bringing out the qualities of the landscape itself. It is critical to note that it is necessary for the cartographer to pay very close attention to the landscape and to understand its processes and ideally to work from personal experience. There is no question here of "automatic" hillshading or whatever. The concepts of "art" and "science" recede in the light of the landscape itself.

SUMMARY

This chapter has suggested that in order to pursue a necessary political project with mapping (cartography and GIS) we need to think about the being of maps at this present moment. This "philosophical" enquiry turns out to have numerous critical outcomes of a practical nature. An important distinction was made between knowledge about things in themselves (what we know) and the horizon of possibility for knowing in general (how we know what we know). It is critical to make this distinction because to fail to do so is to fail to think politically. It is by asking what are the conditions of thinking in cartography that we can both see the shape of that thinking, to see it as it is, and therefore to begin to see how it might be otherwise. It is to think about the being of maps. Seeing this as a critical politics of cartography I suggested a few ways in which this project could be pursued: as a problematization in Foucault's sense (a history and critique of the present), as an ethics, as a struggle, and as a question of technology.

In Part II of this book I shall take up the issue of authenticity and confession through the lens of Foucault's technologies of the self. I shall extend the discussion begun so far by suggesting that our standard understanding of authenticity in cyberspace is mistaken, and that we can better approach cyberspace as the political project of finding our place in the world (where "world" includes cyberspace and physical space). This will require a politics of maps and a politics of ourselves.

PART II

Technologies of the Self

CHAPTER 4

Authenticity and Authentication

The Internet causes billions of images to appear on millions of
computer monitors around the planet. From this galaxy of sight
and sound will the face of Christ emerge and the voice of Christ
be heard? For it is only when His face is seen and His voice heard
that the world will know the glad tidings of our redemption. . . .
Therefore, . . . I dare to summon the whole Church bravely to
cross this new threshold, to put out into the deep of the Net, so
that now as in the past the great engagement of the Gospel and
culture may show to the world "the glory of God on the face of
Christ".

Pope John Paul II, January 2002

As the Pope's World Communications Day message poetically indicates, cyberspace is a difficult issue for the Church. While on the one
hand "the fact that through the Internet people multiply their contacts in
ways hitherto unthinkable open[ing] up wonderful possibilities for
spreading the Gospel", on the other hand the Pope warns that "the
essence of the Internet in fact is that it provides an almost unending flood
of information, much of which passes in a moment. In a culture which
feeds on the ephemeral there can easily be a risk of believing that it is facts
that matter, rather than values" (Vatican Information Service, 2002). For
the Church this raises "necessary questions":

The Internet is certainly a new "forum" understood in the ancient
Roman sense of a public space. . . It was a crowded and bustling
urban space, which both reflected the surrounding culture and
created a culture of its own. This is no less true of cyberspace,

which is as it were a new frontier opening up at the beginning of
this new millennium.

(VIS, January 22, 2002)

Thus, for the Church, as for many others, cyberspace is problematic. It is
ephemeral, fact-driven and empty of values. Yet it could also be the site
of a tremendous and fruitful "opening up". In order to ward off the lack
of values while simultaneously opening up these new possibilities we (or
at least the Church) will have to distinguish between threat and opportunity. Whether as "forum" or "frontier", cyberspace's distant land is dangerous yet attractive, and invites colonization. Insofar as we must cross its
borders we will need to take the right precautions and credentials, and
distinguish between positive and negative, privacy and surveillance, true
from false.

On this view of cyberspace a *crisis of authenticity* arises from a problematic of what comprises "authentic being" (see Chapter 1) and authentic place. This crisis has very meaningful implications for the spatial
politics of cyberspace because it sets into question the true self's place in
the world and the manner in which we come to know authentic place, or
how place is represented cartographically.

The crisis of authenticity has arisen because of a particular and errone-
ous conception of authenticity as a subject's true inner self – an inner
nature or essence which is distinct from what we do in the world. In cyberspace authenticity is constituted as authentication of identity. Where
there is no material evidence (visual recognition, the sound of a voice, the
style of handwriting) we have difficulty in truly identifying people as who
they say they are – in other words, as authentic. Thus, in response to this
crisis of authenticity ever more "confessional" discourses have to be produced. Here, *production* means both producing identifying documentation, for example to the authorities, and production in the sense of making
something new, of creating something, that is, identity in cyberspace.
These identities exist within a regime of normalization (authentic–not
authentic) which are governed and contested through juridical procedures. In this manner, cyberspace is the site of a politics of identity.

In addition, authentication procedures have traditionally been used to
claim an authentic or real existence in physical space, as opposed to an
inauthentic or virtual existence in cyberspace. Whereas self-identification
is presumed to be relatively unproblematic in physical space, cyberspace
is conceived as an especially dangerous place for identity. But these
"authentic places", and "authentic experiences" (ways of being) of physical space, are in themselves constructed in relation to the horizon of possibilities at different time periods; they are historicized and contingent

upon the conditions of the time. In other words, people decide what is "real" by appealing to what is authentic, and when the authentic is cast as what is authenticated, this results in the dividing practices between authentic and inauthentic cyberspace. This has led some to reject the virtual as a non-place, a non-authentic place, or at least to oppose it to physical space. An example of the former rejection lies in the countless difficulties that the field of "digital studies" has had in being taken seriously as a research topic. To "study" cyberspace is somehow not important or real enough to earn tenure, be taken as a serious scholar, or to be seen as "really" doing geography or politics (as opposed to technological work). To counteract such prejudices is of course one of the motivations for this book.

A notable example of thinking of cyberspace oppositionally to physical space is Baudrillard's conception of the simulacrum (Baudrillard, 1994). When Baudrillard asks whether the virtual is real, imagined, a simulation of the real (a modern application of Plato's ideal forms), or a simulation which eclipses the real, he is focusing on the *production of truth about its being*. We may either denigrate or glorify the truth of this being, but in doing so our judgements concerning its authenticity rest fundamentally on what kind of truth it can produce.

However, the notion of authenticity as authentication does not exhaust the possible set of meanings of authenticity. In this chapter I address the cyberspatial problematic of authenticity and suggest a different conception of authenticity as the practice of working on the self as finding our place in the world. Authentic selves and authentic places are related in the sense that an authentic care of the self is concerned with how we are "in place". This in turn leads to a different conception of confession. Therefore in the next chapter I shall offer a critique of confession and examine some techniques of the self such as self-writing and *parrhesia* which might offer a more positive way forward. As I shall describe, authenticity and confession constitute a "nexus" of relations in cyberspace. Their division here into two chapters is not meant to imply they can be cleanly broken apart.

AUTHENTICITY AS AUTHENTICATION

As the Pope's message makes clear, authentication of identity and place is notoriously difficult for cyberspace. Indeed, as Turkle (1995) argued, identity is fluid. But it is *too* fluid for many. What is needed to sort out the problem of cyberspatial identity are procedures for authenticating valid email, for checking people logging on, for tracing the origin of communications across networks using email headers, for safely perform-ing e-commerce and verifying credit card ownership, for being a "valid" group member, and for checking chat room identity, just to name a few of the authentication procedures now routinely employed. These proce-dures are not always successful. Indeed, much discourse about authentic-ity as authentication revolves around the need for *further* procedures, or how to meet the constantly growing difficulties of separating proper access from improper access (e.g., hacking) or how to grant access to different groups of people on the campus server (e.g., students, staff, faculty), that is, to again differentiate between the identity of different groups of people. As methods become more sophisticated for "spoofing" email or capturing passwords so there is required an equally sophisticated set of methods to determine the real identity from the fake.

An example may help clarify this concern. The following are remarks from an introduction to a network encryption protocol:

> **The Internet is an insecure place**. Many of the protocols used in the Internet do not provide any security. Tools to "sniff" passwords off of the network are in common use by malicious hackers. Thus, applications which send an unencrypted password over the network are extremely vulnerable. Worse yet, other client/server applications rely on the client program to be "honest" about the identity of the user who is using it. Other applications rely on the client to restrict its activities to those which it is allowed to do, with no other enforcement by the server.
>   (Source: http://web.mit.edu/kerberos/www/)

In this formulation the Internet is characterized as a special, even separ-ate, place with the property of dangerousness (it is insecure and unsafe). This dangerousness is due to a possibility of threats to real or true iden-tity from bad or hurtful forces ("malicious hackers"). These forces may be able to interrupt the generation of the truth about oneself, the "honest" (true) confession of identity which needs to be transmitted to others. Identity is a matter of being able to satisfactorily prove to others that you are who you say you are. Authentication is a matter of being authorized by authorities on the matter of your truthful confession. In cyberspace, for technical and social reasons this kind of authenticity is more difficult to achieve than in physical space, where we are thought to have other lines of proof (visual appearance, sound of one's voice, finger-prints, DNA, etc.). Concern over truthfully establishing one's online identity has magnified over the last decade as the storage, processing, and analysis of personal data has increased. In effect, more and more of our personal information is stored online so that we almost have a second identity (Lyon, 1994, calls this our "data-image") which can override our physical presence.

One of the first people to bring to popular attention the problems with data-images was David Burnham in his appropriately timed book on the rise of the computer state (1984).[1] Burnham's book was stuffed with examples of individuals who had suffered because of errors in official and commercial databases which trashed their credit ratings, mistakenly listed them as AWOL from the military, and privileged computer information over other knowledge. Rereading this literature today gives one a strange sense of déjà vu, as well as surprise at just how far things have moved on. As late as 1990, for instance, the privacy threat of Caller ID was still an issue! It is hard to imagine that just over ten years ago people were actively opposed to Caller ID because of its potential for privacy loss because today it is so normal.

A dramatic illustration of these concerns was *The Net*, a mid-1990s film starring Sandra Bullock. This film articulated precisely the concerns of a society fearing loss of identity and even substitution of a damaging false or non-real identity. As the main character says:

> Just think about it. Our whole world is sitting there on a
> computer. It's in the computer, everything: your, your DMV
> records, your, your social security, your credit cards, your medical
> records. It's all right there. Everyone is stored in there. It's like
> this little electronic shadow on each and everyone of us, just, just
> begging for someone to screw with, and you know what? They've
> done it to me, and you know what? They're gonna do it to you.

The movie was advertised under the tagline "Her driver's license. Her credit cards. Her bank accounts. Her identity. Deleted". Identity and self are a "shadow" of some true inner being which can be made subservient to the digital records which provide authentication. In Paretsky's novel *Indemnity Only* a character makes a slip-up between authorization in the sense of permission and authentication of selfhood (see box). In both *The Net* and *Indemnity Only* the self is static and stable and not historicized or part of the world. But this notion of authenticity is far from the sense of the care of the self which we will explore later.

A more significant example might be the announcement in 2001 by several leading publishers of medical journals (Kluwer, Blackwell, Springer-Verlag, John Wiley, Elsevier, and Harcourt) that they will make available their journals to medical schools in poor countries, with funding from WHO (Brown, 2001). Given the inequalities of knowledge

---

Indemnity Only

Ajax occupied all sixty floors of a modern glass-and-steel skyscraper. I'd always considered it one of the ugliest buildings downtown from the outside. The lower lobby was drab, and nothing about the interior made me want to reverse my first impression. The guard here was more aggressive than the one at the bank, and refused to let me in without a security pass. I told him I had an appointment with Peter Thayer and asked what floor he was on.

"Not so fast, lady", he snarled. "We call up, and *if* the gentleman is here, he'll authorize you".

"Authorize me? You mean he'll authorize my entry. He doesn't have any authority over my existence".

(Paretsky, 1982, pp. 23–4)

---

distribution between developed and developing countries this example seems practical and humanitarian. Access will be electronic, through a special WHO Web portal. Similarly, many university libraries now offer online full-text access to peer-reviewed journals for students and staff (EBSCO, OCLC, etc.). But on closer inspection we see a rather interesting qualification – access to the medical journals is governed by the per capita income in a country: where per capita GNP is below $1,000 the journals are free, between $1,000 and $3,000 there is a minimal charge, therefore authentication here reaches out back to physical space in order to verify something. This shows how, once again, cyberspace and physical space are in a constant relation of mutual production.

The "slip" from authorizing a particular behavior to authorizing one's existence is all too easy to make. Throughout cyberspace, the procedures for authorizing entry are numerous and frequent, beginning with one's "initial" entry (although again such language can dangerously ratify cyberspace as objectified space). Depending on the degree to which you can confess, you are permitted different degrees of access. Modern networked computers (e.g., Windows NT or derivatives thereof) present a dialog box during logon which asks for a password. In some instances one can bypass the logon or enter a different (but valid and authentic) password. If one does this one enters a different relationship with the computer. One possibility is to enter as a "default" user where individualized programs and settings are not available. Another possibility is to enter as a "super-user" or "administrator" where one has "privileges" which allow changing the system settings, accessing other computers over the network, or installing software. The degree of identity authentication is finely tuned with the degree of confession.

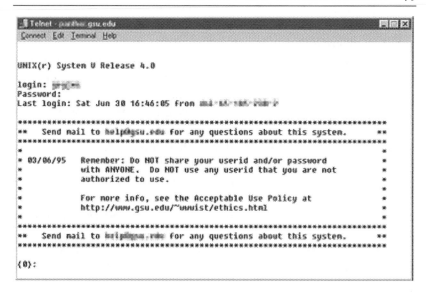

Figure 4.1   A typical logon screen demanding password, and warning against "confessing" the password to someone other than the appropriate authorities, as well as illegitimately attempting to authorize one's identity. Used with permission.

Let's look at an actual logon screen (see Figure 4.1). There are several procedures for ratifying oneself here. First, one can have one's identity sufficiently authenticated through the confession of the correct password. The confession shows that one is able to produce the truth, in this case the truth about oneself (that one is oneself, and not, say, another, inauthentic being). But this confession must take place within certain strict guidelines; one cannot confess the password repeatedly and openly or to just anybody: "Do NOT share your userid". Only a confession to the appropriate authorities can produce the truth. Indeed, if one did confess one's password to someone other than the correct authority, this password would by definition no longer be the truth, because the truth can only exist in a relation between subject and authority. Another password would have to be issued by the authorities. A second injunction occurs against those who are inauthentic and try to use the same strategies as the authentic to produce the truth: "Do NOT use any userid that *you are not authorized* to use" (emphasis added). You are not authorized. Authenticity here is authentication to cross the border into cyberspace – to be identical to your self (and not another) upon which all further interpretation of who you are will be founded.[2] So there are several injunctions and procedures at work here which rely on an ability of the truly authentic to produce the truth about themselves. This paradoxical description of the "truly authentic" thus raises a central problem for cyberspace, one which to be sure is also found in physical space, but in a slightly different way.

Take the example of a simple housekey. Possessing the key allows one to enter the house as long as one has the right (authentic) key. A housekey can be shared, copied, or leant out legitimately. In fact, possession of the key might designate a legitimate reason for entering the house (as when the key is leant out to a short-term guest who can produce the key if challenged by a neighbor to legitimate their entry). If challenged by the police, one could provide the key as support for one's being in the house. Or, putting this into the way we have been thinking about this so far, the *confession produces a truth which authorizes one*'s *being* in the house. However, possession of a password in like manner to possessing a housekey is strictly forbidden; only one person can possess and confess the password to the authorities – sharing of computer accounts is forbidden as a condition of use. This tighter rein on the confession–authentication nexus indicates a greater concern in cyberspace over identity.

It is interesting to reflect that the authorization is over one's being (in the house, or online). We saw this just now with the Paretsky quote. Authorization in the sense of permission can easily slip to authorization in the sense of governing one's being and identity. This is a very dangerous slippery slope which is not easy to avoid when authenticity is authentication. We shall have to see if there are ways of avoiding the slope if we are to have an authentic existence in cyberspace.

But these little examples are only the most simple and obvious strategy of authentication that a person may encounter. This does not mean that there are not many that are occurring below the surface, instances when identifying identity becomes a highly political matter. Consider the case which occurred during the G8 summit in Genoa, Italy, during the summer of 2001. A network of independent journalists known as Indymedia published a report that Italian police were attempting to "infiltrate" their IRC chat forums. The IRC, or Internet Relay Chat, is an Internet-wide standard for live chats which take place on different channels. Users can log in and chat to one another in real time. Indymedia reported that the Italian police logged in under an assumed nickname ("crudelia"), it is assumed, in order to listen to plans and coverage. One Indymedia member was quoted anonymously as saying "We know their IP addresses and hostnames . . . When we see them come on, we kick them off. Unfortunately, we are really busy during major actions and cannot spend all of our time monitoring police spies". The article goes on: "The activists report that police often come onto their server, pretending to be fascists or extreme leftists advocating violence. Italian activists have assembled their own form of counter-intelligence, keeping track of hostnames and other clues which can help them identify cyber-cops" (Indymedia, 2001).

The Indymedia example, as well as the other examples cited, indicate some of the political ramifications which necessarily occur when authenticity is implemented as authentication. Immediately at issue is identity, confirming, allowing, or denying it. For some, such as Turkle, the Internet and cyberspace provide almost unparalleled opportunities to play with identity, to construct it in interesting ways which might resist or even break out of hegemonic power structures of the physical world. But we must now see this as only half of the picture. When identity is problematic we invent an equally impressive array of procedures for determining identity, for fixing, stabilizing and normalizing it. We are then "back" in power relations. If, as Foucault said, there is no power without resistance, it would appear that there is also no resistance without power. There are few new answers to online subjectivity in identity-play. The crisis of identity-authenticity as authentication is a turning away from what could be an authentic relationship with, through, or in cyberspace. An authentic existence as the care of the self as finding our place in the world.

WHAT SPACE FOR AUTHENTICITY?

> . . . then [the individual] was authenticated by the discourse of
> truth he was able or obliged to pronounce concerning himself.
> Foucault, *History of Sexuality, Vol. I* (1978, p. 58)

Having now considered some examples of authenticity as authentication in cyberspace we can explore a deeper set of possible meanings for authenticity. These meanings will lead us to consider how authentication and confession can be reinterpreted as technologies (or techniques) of the self. That is, how subjects produce themselves in cyberspace, and how they are at the same time produced in a set of mutual relations.

In order to answer this question we shall consider how authenticity was discussed by Heidegger in his book *Being and Time*. Heidegger's phenomenological work has had considerable impact on thinking about space and place. In many ways his highly original thinking about what it means to be human by highlighting being (as in "being" human), and the centrality he afforded to our being-in-the world, brought space in right from the start. What is the spatial essence of this world, and how are we "in" it? Are we really in the world in the sense that the water is in the jug, or is there some other, more authentic way we are oriented with the world? How are we with others (e.g., can we be with others at a distance, such as over the Internet)? All of these are spatial questions, as Heidegger knew (see §§22–4).

For Heidegger, authenticity is "being at home with oneself" or "self possession". When we are in possession of ourselves we face up honestly to our situation in the world. In typical fashion Heidegger looks at the etymological connections between these senses and underlines a relationship between "authentic" [*Eigentlich*] and having oneself as one's "own" [*eigen*] (Heidegger, 1962, p. 68, fn. 3). He notes:

> And because Dasein [human being] is in each case essentially its own possibility, it can, in its very Being, "choose" itself and win itself; it can also lose itself and never win itself; or only "seem" to do so. But in so far as it is essentially something which can be *authentic* – that is, something of its own – can it have lost itself and not yet won itself.
>
> (1962, p. 68)

This means that Dasein is "mine to be in one way or another" (Heidegger, 1967, p. 68) and can be lost or embraced. If Dasein is lost then it is inauthentic, but that does not make Dasein "less" or "lower", but rather where it has become more concerned with the world rather than with the question of its Being: "it is the case that even in its fullest concretion Dasein can be characterized by inauthenticity" (p. 68). Thus inauthenticity is not in and of itself a lesser kind of being. This sheds some interesting light over the authentication–confession nexus we have described until now, because the nexus has implied that indeed inauthentic being *is* of a lower or lesser kind. It is lower because it has not been able to verify its identity by producing (confessing) the truth about itself. If we cannot produce the right password (or housekey) then we have fewer rights of access. On Heidegger's suggestion, however, even a truly inauthentic being does not prevent Being from reaching its "fullest concretion".[3] In fact for Heidegger inauthenticity most likely occurs when we are busy paying attention to everyday life and fully engaged with our projects themselves, for example when we are excited or daydreaming. This is far from the deprivileged identity associated with identity–authentication.

When authenticity is determined by authentication, we pass over the possibility that people can be fully, even essentially and "genuinely" inauthentic, at least for particular periods of time. Therefore we fail to think very much about the qualities of being inauthentic and how that inauthenticity may become authentic. We also dismiss the inauthentic as nongenuine and less worthy of attention. But such inauthentic moments frankly may outnumber the authentic, and not to pay attention to it may be to cut ourselves off from how much of the world "really" is. For example, the geographer Edward Relph made a now-classic distinction between "place" and "placelessness" as one of authentic or inauthentic places. Relph allied the authentic with the genuine and the inauthentic with the artificial: authenticity is "that which is genuine, unadulterated, without hypocrisy, and honest to itself, not just in terms of superficial characteristics, but at depth" (Relph, quoted in Peet, 1998, p. 50). For Relph industrialized mass society with its telecommunications, mass media, and big business is inauthentic placelessness. In contrast to Heidegger, who thought that people could be genuine in their inauthenticity, and actually were so most of the time, Relph casts it as hypocritical and not honest to itself. It divorces us from a true relationship with real places.

On Relph's account the Internet is distinctly placeless. It is artificial, arbitrary, and seemingly "no-place" (utopia). It has no depth, it is all superficial surface without any sense of place. But on Heidegger's account Relph's conception of "place" turns out to be too narrow; there are genuinely inauthentic places in the world which we need to take into account. Cyberspace is not a separate "other" placeless geography distinct from the authentic physical world. It is part of the place we find for ourselves in the world – it is everyday life – and is no more or less authentic than the rest of it.

Of course, Heidegger does value authenticity and even considers ways in which we might be "fallen" from it through being determined by the mass of others (see §§35–8). This fall from authenticity amounts to letting other people direct or "own" our ways of being instead of doing it ourselves. Thus it is when we attend to or take care of our selves *and* our place in the world that we are authentic. In contradistinction then to the conception of authenticity as truly identifying a subject, of nailing down a stable identity, authenticity can now be better interpreted as exercising the practices of oneself. It is a question of care (of the self) as how it finds its place in the world.

TECHNOLOGIES OF THE SELF

The problematic of the care of the self reaches its most mature and influential position in the work of Foucault. This is not the place to compare the thinking of Heidegger and Foucault on this issue, but it is interesting that there is today an increasing realization that Foucault often worked in the light of Heidegger, fruitfully reconceptualizing and reappropriating (Elden, 2001a). As I shall discuss in the next chapter, it was Foucault's insight to show that confession is produced as a normalizing practice. At the same time further work (e.g., Burchill, Gordon and Miller, 1991; McWhorter, 1999; Rose, 1996) has profitably extended Foucault's formulations more generally into what has come to be known as "governmentality". It is still a task to investigate spatial politics under this heading of governmentality (see, for example, Hannah, 2000).

In order to further develop our critique of cyberspace, we need to examine some of Foucault's ideas on the technologies of the self. They constitute a theme Foucault was working on from about 1978. Broadly speaking it means:

1. a set of techniques and practices that can be deployed to modify or affect the self;
2. these techniques are historically situated within power relations (historicized);
3. they can mix one's own knowledge with that of the rule (the "contact point" of governmentality mentioned in Chapter 1);
4. as with other writing in Foucault about power, it has a very productive side (producing the self for example) as well as a constraining side.

By using the word "technology" Foucault meant a somewhat wider use of the term than is usual in everyday language; the Greek *technē* [τεχνη], production or producing something new, a fabrication. Heidegger says "*technē* is the name not only for the activities and skills of the craftsman, but also for the arts of the mind and the fine arts. *Technē* belongs to bringing-forth, to *poiēsis*; it is something *poietic*" (Heidegger, 1977, p. 13). By "bringing-forth" Heidegger alludes to his conception of how the truth is made present in people's lives; not in the sense of it being hidden, lurking, and in need of recovery, but just of being brought out ("opened up") and looking beyond what is present-at-hand (Heidegger, 2000, p. 169). *Technē* is a reference to this ability to open up the truth. Heidegger points out that in Plato's *Phaedrus* (260d–274b) the discussion revolves around the question of how rhetoric can become a proper *technē*, which is "generating, building, as a knowing pro-ducing" (Heidegger, 2000, p. 18).

Also, in Aristotle, we find the following usage:

> Now architectural skill, for instance, is an art [*technē*], and it is also
> a rational quality concerned with making; nor is there any art
> which is not a rational quality concerned with making, nor any
> such quality which is not an art. It follows that an art is the same
> thing as a rational quality, concerned with making, that reasons
> truly. All Art deals with bringing some thing into existence; and to
> pursue an art means to study how to bring into existence a thing
> which may either exist or not, and the efficient cause of which lies

in the maker and not in the thing made; for Art does not deal with
things that exist or come into existence of necessity, or according
to nature, since these have their efficient cause in themselves.

(Nicomachaean Ethics, 1140a 1)

This passage is useful because it emphasizes how an art, a *technē*, can
bring something into existence, and what is brought into existence in
technologies of the self is precisely the truth about oneself. And this
becoming into existence is a production, it does not occur "of necessity
or according to nature", says Aristotle, but is a technique that has been
adopted. Nature does not need a *technē*, but we (we humans) do, in order
to "bring into existence" ourselves.

One might produce something entirely from scratch or one might
shape and mold something from nature, one might even imitate (*mimetike*), but in all cases something has been produced (cf. Sallis, 1999, p. 16).
In speaking of technologies of the self therefore, Foucault is concerned
with those techniques which mold and shape, which *produce* the self. This
self-production occurs throughout the lifespan and is quite distinct from
the modern idea that a self has to be recovered or exfoliated. It explains
Foucault's remark that one must work "to create a gay life. To *become*
[gay]" (Foucault, 1997f, p. 137) for example.[4] Obviously the political ramifications of such statements are near the surface here, indeed even at
odds with much thinking about the politics of homosexuality (although
not all – see McWhorter, 1999; Warner, 1999). The politics of care of the
self will be considered at more length in my concluding chapter, but for
now we may note an important passage in Plato's "Apology" (29d–e)
where Socrates points out to his judges the discrepancy between acquiring wealth and honors, living in a city renowned for wisdom and power,
and their lack of care for truth and the best possible state of their souls.
Throughout his defense Socrates presents himself as the honest "gadfly"
whose activities are important for the city (*polis*) if it is to be a city. So he
says after they find him guilty:

> I tried to persuade each of you not to care for any of his
> possessions rather than care for himself, striving for the utmost
> excellence and understanding; and not to care for our city's
> possessions rather than for the city itself.
>
> (Pl. *Apol.* 36c)

Here then the care of the self helps to produce the *polis* as the political,
the way the city should be run or governed. The self and the political are
conjoined.

Foucault found that techniques of the self vary at different times. In one general distinction that he worked on during the early 1980s (including his 1980 Cours on "The Government of the Living" in Rabinow, 1997) he contrasted the Greek work of Stoics such as Seneca and Epictetus who urged that one perform regular stock-takings of oneself, including a process of self-writing called *hupomnēmata*, and that of the early Christians who developed a series of *confessional* techniques where one is only authenticated by revealing oneself to a master or superior (Foucault, 1997c, p. 209ff.). Another part of this same contrast was that the Christians tended to believe that one was reaching a true self hidden inside oneself, while the Greeks were more interested in techniques which could transform yourself. For these Greeks, you had autonomy even if you were a pupil to the master; you sought him out not be constantly confessing your sins, but to help you develop yourself. Foucault did not hesitate to see the Christian confession as resulting in a renouncement of oneself (at its height the martyrdom advocated by writers such as Tertullian and Jerome, where a particular aspect of confession, called *exomologesis* demonstrated "a way for the sinner to express his will to get free from this world, to *get rid of his own body*, to destroy his own flesh and get access to a new spiritual life . . . it is the dramatic manifestation of the renunciation to oneself" – see Foucault, 1999, p. 173, emphasis added). In the next chapter I shall argue that this very same impulse – to shed the body in the confessional mode – is a problematic of cyberspace.

Foucault distinguishes four aspects that the relationship to oneself can take on. These aspects can be used to analyze the production of the subject "in" cyberspace. They all have to do with producing the truth about oneself, in other words of more authentically becoming (where the process itself, becoming, is what is important). The first is concerned with morality, such as feelings, intention, and desires. As far as our contemporary society is concerned, it is feelings which are paramount in the moral equation, contrasting with that of the Christian where it is desire which is problematic and has to be dealt with. This he calls the "ethical substance" (Foucault, 1997b, p. 263). Second, "subjectification" [*mode d'assujettissement*] or how moral obligation is produced, "how people are invited or incited to recognize their moral obligations" (p. 264). What is it that causes people to take up the issue of ethics – is it the law (natural or divine), a rationality, or the imperative of living beautifully? Third, are the self-forming activities, what we have to do, which he called an *asceticism* in its broadest sense, a "training" (Foucault, 2001, p. 143 ff.). Fourth, and finally, in this set of possible aspects of technologies of the self, is the teleology or goal we aspire to. Depending on this goal there will be different practices or *askēsis*. For example, if you want purity of being (the kind

Table 4.1 Technologies of the self

Society	Aspect of morality (ethical substance)	Subjectification (mode d'assujettissement)	Practices of the Self (form d'ascese)	Teleology of Ethics (goal of being ethical)
Greeks – Classical	Acts – pleasure – desire (aphrodisia)	Beautiful existence	Technē for your role	Self-mastery in order to rule
Greeks – Stoic	Acts – pleasure – [desire] aphrodisia	Rationality	Exercises in restraint	Self-mastery to be rational
Christians	[Desire, flesh, body]. To be eradicated	Religious, juridical (divine law)	Techniques of self for self-decipherment (confession)	Immortality, purity
Middle Ages	Desire, –?		Techniques of self examination	Cure of souls?
Present	Desire, feelings	Juridicality Discipline Power–knowledge	Confession	Liberation from power, Production of truth
New – present or cyberspace	Authentic care of the self	Contact point of governmentality	Practices of pleasure	Maximizing freedoms within power relations

Source: Foucault (1997b, pp. 263–9; 1985, pp. 25–32).

of ethics found in early Christianity), you will go the confessional route, whereas if you want self-mastery in order to rule others (the kind of ethics found in Plato) you will perform practices which correspond to the beautiful life (clean manliness or *aretē* [αρετη] for example).

In order to contrast these four different aspects of the technologies of the self and their varying implementation in different time periods, I have constructed Table 4.1 from Foucault's remarks. All rows (except the last, which is my own formulation) are derived from descriptions in Foucault's work.

Foucault uses this idea of the technologies to claim that he is doing a theoretical analysis with a distinct political dimension (1999, p. 161). By politics he meant "what we are willing to accept in our world, to accept, to refuse, and to change, both in ourselves and in our circumstances" (p. 161). This is perhaps an unusual usage of the word, but it corresponds to his final statement in these lectures, which I think is very important to keep in mind, namely that he was seeking the "politics of ourselves" (p. 181). Technologies of the self then, although seemingly delving into the heart of classical philosophy, are the site and foundation for political intervention in the real world. They are part of an approach melding theory and practice. In this respect, the second two volumes of the *History of Sexuality* are all too often overlooked in favor of his more popular works. I would argue that this is mistaken, and that they constitute the framework for a political history of the subject. This political framework has direct bearing on our interpretation of cyberspace because it foregrounds the question of what practices of the self in cyberspace can maximize our freedoms within a disciplinary regime (see last row of Table 4.1).

Foucault is often associated with the negative effects of power. However, in his later work he admitted that his earlier work had dwelt too much on technologies of domination. He now wished to correct that by an adjustment which would bring into focus technologies of the self. But this should not be an over-adjustment, where we swing too far the other way to a completely self-determining subject. One should see that power is complex, that "relations involve a set of rational techniques, and the efficiency of those techniques is due to a subtle integration of coercion-technologies [the disciplinary society] and self-technologies" (1999, pp. 162–3). The upshot is that the subject is the contact point of governmentality.

We see then how authenticity comes to mean how truth comes to be truthful, or how truth achieves its authenticity. (1) We need to know ourselves in order to assess how well we have achieved our goals, lived the beautiful life, or behaved rationally (Table 4.1). (2) This requires producing the truth (through a self-examination for the Stoics or confession for the Christians) through technologies of the self. (3) The truths that are produced are assessed, or authenticated, and placed within a regime of normalization. In the next section I will provide some evidence of how this conception of authenticity plays out in cyberspace, before going on to explicate the final row of Table 4.1 which recasts technologies of the self as practices of pleasure to maximize freedoms.

AUTHENTICITY OF PLACE AS A POLITICAL PROJECT: AGAINST THE "CONFESSION" OF THE MAP

The question of spatiality has never been far away from our discussion of authenticity. It is that process in which we as humans can fruitfully practice care of the self finding its place in the world. How are we spatially attuned to the world? Space or place in cyberspace should not be assessed on the grounds of how "well" we can fit our experiences and projects "into" it (e.g., by asking how we can have an authentic experience in cyberspace), but rather to what extent it provides horizons of possibilities for fitting with our projects.[5]

What are these projects as far as the authenticity of cyberspace might be concerned? This question can be pursued as a practice of the self, whether in cyberspace or physical space or some conjunction of the two. There are no doubt many such possible practices and not everyone will want or be able to pursue them all. In my own case I suggest that mapping is an integral part of such finding one's place in cyberspace and more generally.

As I discussed earlier, the issue of our mapping of cyberspace is a political project. Mapping – making maps and representing the landscape – is a significant and irreducible activity in the production of geographic knowledges, whether for physical or virtual spaces. But perhaps because cyberspace is not perceptible in the same way as physical spaces and places we have come to rely quite heavily on mapping cyberspace as our cyberspatial imaginary (e.g., see Dodge and Kitchin, 2001). Maps make the abstract physical. They help us visualize something we cannot see. But they also construct ways of thinking. I have argued that modern mapping largely operates as ontic knowledge, that is as knowledge of things in themselves, as opposed to widening the condition of possibilities for ontic knowledge (ontological knowledge). Now I wish to show that the current practices of mapping cyberspace are constituted as technologies which "confess" the truth of the landscape.

For a long time now geographers have assumed a model of mapping which requires belief in the fact that the map "confesses" the landscape. This confession is both coaxed out and blocked from fully emerging. It is coaxed by the authoritative expert – by the use of better map symbols which communicate their intent more clearly, by user research in map recall and interpretation (i.e., the field of "cognitive cartography"), but ultimately blocked by the being of the map. The map after all is generalized, at a reduced scale, reprojected from the sphere, and suffers misinterpretation by users. The landscape is never fully present in the map, but instead deferred and repressed. If only we could entice the landscape to fully emerge! Would this not lead to more accurate mapping, more useful maps? This thought has lead to a whole deployment or apparatus (*dispositifs* as Foucault called them in *History of Sexuality*) of techniques for measuring and releasing more accurately the subject of the map's confession: the land itself. These techniques take place within a great field of scientificity, beginning with the attempted application of the information signal processing work of Claude Shannon after World War II which finally culminated in the so-called Map Communication Model (MCM). Communication is still the sign under which mapping and GIS are read (Goodchild, 2000). This communication is but confession under a new name.

In Chapter 3, I discussed the work of Harley, arguing that he pushed forward the problematics of spatial representation, but that at the same time we need to move further than Harley had time to do (see also Crampton, 2001a). As I have tried to show, Harley's theorizing of the map ultimately came down to a representationalist model, one where the landscape would appear in the map, if it were not for ideology. Clear through the ideology, Harley says, and we may release the full confession of the map. This was a very valuable move because it focused on the social and political relations of the map for the first time. In Foucauldian terms the map became problematized as a relation of power-knowledge. But it was not a sufficient one for a critical cartography. In sum, it provided an epistemological avenue into the map, but still left open the question of the ontology of the map.

We can pick up this point by considering some of Harley's later work on the silencing effects of the map and its role in colonialization. The year he died, 1991, Harley was preparing a major exhibition of mapping in association with the 600th centenary of Columbus' voyage to the West Indies in 1492. For Harley, who was already almost a decade into the ground-breaking work of a complete history of cartography with David Woodward, this provided a test case for some of his theoretical work. It was not to be a celebration of Columbus, but rather a cartographic documentation of an "encounter" between an already existing population and newcomers. In an article published posthumously, Harley (1992) was quite frank about the power effects of Columbus' voyages, for example the renaming of places and the loss of the original names (one or two were recorded by Columbus in his log, such as Guanahani, which he renamed San Salvador as part of Columbus' avowed "Christianizing mission").

This renaming of original places had been previously discussed in one of Harley's better known pieces on "secrecies and silences" of mapping (Harley, 1988b). This covered a wider range of examples, including modern military gaps and white spaces on the maps where bases are located, as well as the renaming, or now, as he understood it, the silencing of native voices. This silencing of the landscape was part of a series of strategies present in mapping which had to be dug out from their hidden depths, like a confession is dug out of the penitent. For example, Harley says "silences should be regarded as positive statements and not merely as passive gaps" (1988b, p. 58). The map therefore is a domain of buried truths which a strategy of confessional techniques will release. This was not so much ideology as false consciousness, although he may have had something like this in mind with his projected book on the "ideology of maps" (with John Pickles) which never appeared.[6]

The lesson we take from Harley then is not that maps cannot represent the landscape, but that their attempts to do so are ringed around with biases and ideology. Nowhere does he say or imply an otherwise condition, where maps are not ideological, and I have no doubt that he did not think it was worth even pursuing that route. Instead, our task as critical cartographers and geographers was to uncover the ideologies for particular sets of maps. That way we could at least unburden the load they carry, if not quite dig out the secret truth residing in their core.

For these reasons I think Harley's work stopped at the fatal step. He was able to show the power-knowledge relations of mapping, but unlike Foucault, he did not abandon the need to uncover (confess) the foundationalist origin. As far as representation goes then, Harley and Goodchild have the same theory.

SELF-WRITING AS NON-CONFESSIONAL PRACTICE OF THE SELF

Foucault suggests an interesting aspect of Greek practice which is relevant to our rethinking of authenticity – the technique of *hupomnēmata* or self-writing (Foucault, 1997b, 1997c, 1999). I would like to end this chapter on authenticity by briefly examining what this could mean for our relationship with cyberspace. It has vibrantly rich potential for a non-confessional practice of the self.

The Greek technique of *hupomnēmata* consisted of notebooks or aide-memoires in which one could write down examples and reflections, in order to work on the self, and to produce the truth about oneself. The standard dictionary of Greek defines *hupomnēma* as "reminders, memorials, notes, or a memorandum" (literally it means "under [the influence of] a memorial or record").

Foucault suggests that they were first introduced in Plato's time for personal and administrative use (Foucault, 1997b, p. 272). For example, in Plato's *Phaedrus*, we see Socrates speaking of the commendable practice of writing, not to lock it into place but rather, like a sensible man of husbandry:

> The gardens of letters he will, it seems, plant for amusement, and will write, when he writes, to treasure up reminders [*hupomnêmata*] for himself, when he comes to the forgetfulness of old age, and for others who follow the same path, and he will be pleased when he sees them putting forth tender leaves.
>
> (Pl. *Phdr.* 276d)

In other words, by writing down things learned, one can create a "treasure" constituting "reminders" which one might need to call upon when one's memory is weak, as may occur in old age. Socrates likens this to planting seeds or "gardens of letters" which will be productive or "sprout new leaves". What is produced in these self-writings is the truth. But they were not simply reminders to be used when memory is weak, but the framework for practice itself. They should be *ad manum* or "near at hand" in order to be used whenever they were needed, so near in fact that they should be "planted in the soul" (Seneca), "in short, the soul must make them not merely its own but itself" (Foucault, 1997c, p. 210). Self-writing, then, *is* ourselves in our being.

This innovative use of copybooks was a new technology, which "was as disrupting as the introduction of the computer into private life today" (Foucault, 1997b, p. 272). Through this practice, one can administer one's knowledge, or remind oneself of useful things that one has heard. Thus they "give access to the truth" (p. 272) through a "communion with memory" (Pl. *Phdr.* 249c).

By the time of the later Greeks, especially the Stoics, Foucault suggests that they now constituted a practice (*askēsis*) of a form of virtue, to exercise mastery over oneself in the art of living (*tekhne tou biou*) in order to take care of oneself (in Latin *cura sui*, or Greek *epimeleia heautou*). The word "ascetic" has accrued some modern meanings which tend to overlay the original Greek meaning of practice. As far as Diogenes goes, for example (often called an "ascetic"), Sloterdijk argues that "[w]e have to rid the word of its Christian connotations to rediscover its fundamental meaning. As free of need as Diogenes appears, he could be taken rather as the original father of the idea of self-help" (1987, p. 158). Here Sloterdijk is very close to Foucault's notion of the care of the self. In the sense which we shall be later using, *askēsis* comprises a way of life as an ethics, a set of mores, practices. This important and more originary sense of ethics can be found in both Foucault and Heidegger (see Chapter 8).

If one did not practice or exercise constantly, in this case through writing, then one could not properly constitute oneself. By contrast with *anamnesis* or remembering, the *hupomnēmata* were not an attempt to recover what was forgotten, despite being notebooks for memory. They are not confessional in nature. Rather, they had the aim of recording notable things one had heard or read "to make one's recollection of the fragmentary *logos*, transmitted through teaching, listening, or reading, a means of establishing a relationship with oneself" (Foucault, 1997c, p. 211).

The sense of being near or present at hand alluded to above could also work with another kind of writing, that of correspondence. Here the sense is not of being present to oneself, but to others, even when there is physical distance between the writer and the recipient. By writing a letter, the writer becomes present to the one who receives it "with a kind of immediate, almost physical presence" (Foucault, 1997c, p. 216). Writing is an activity of showing yourself, of projecting oneself into view, one offers oneself to be seen, "in a sense, the letter sets up a face-to-face meeting" (p. 216). Here we see Foucault in a particularly Heideggerian moment of talking about the self as mutually being in each other's presence ("being-with" as Heidegger would say) and "opening oneself up to others" (1997c, p. 217) by "projecting oneself" (p. 216). These notions of being-with each other even if physically distant may be reinterpreted in the context of cyberspace. They suggest that such things as authentic community-formation may be possible through an *askēsis* of self-writing (and possibly other practices). This is not so much the death of distance, or the annihilation of space, but rather the production of community in (cyber)space. This is a radical departure from the conception of authenticity as authentication. It suggests that an authentic finding one's place in the world through practices of care is a distinct possibility. What this authentic community-formation may look like in cyberspace is the topic of the next chapter.

CHAPTER 5

Communities in Cyberspace: Confession and *Parrhesia*

The Internet is a wonderful instrument for evangelisation and pastoral service, but it will never be possible to confess online.
Archbishop John Foley, the president of the special council, the Vatican, 2001[1]

At one point, fed up with civil mechanisms, Alice tells her lawyer: "Let Oprah be the judge . . . Let Robbie and me, Mrs. Mackessy, Howard, Theresa, Dan, Mrs. Glevitch – let all of us come before Oprah. Let the studio audience decide. They're nice suburban woman, many of them, dressed for a lark. They have common sense and speak their minds". Apparently La Winfrey was listening, since she chose this beautifully observed novel for her book club.
From Amazon.com's review of *A Map of the World*, by Jane Hamilton[2]

ON SPEAKING YOUR MIND

In this chapter I examine the competing roles of confession and *parrhesia* (frank speech) in the production of the self in cyberspace. In particular I wish to examine, if only in a preliminary manner, the way that cyberspatial technologies extend, develop, distribute, and reproduce the practices of confession found "in real life", that is in the physical world (again, remembering the problematics of this distinction which were discussed earlier). As we saw in the previous chapter, the relationship between confession and authentication is mutual. One confesses in order to be authenticated. And authentication requires confession. When there is a "crisis" in authentication, as there seems to be with cyberspace, then ever more confessional practices are called for and produced.

One way to challenge this prevailing view is to reconceptualize cyberspace spaces such as chat rooms, blogs, and online diaries not as "confessional", but as "self-writing" techniques of the self in a community or world. Like the sacrament of confession, in self-writing techniques one enters into a relationship with oneself; unlike juridico-religious confessions, however, they do not have the goal of releasing some inner truth buried within oneself but rather of *working on oneself in the context of a community (world)*.

BLOGGING AND COMMUNITY

The recent rise of "blogging" is such a case of working on oneself as the practice of finding oneself in the world. *Wired Magazine* recently called weblogs "the hottest publishing phenomenon on the Internet" (Kahney, 2001, n.p.). Used in the sense of a weblog or frequently updated journal or diary kept online, I have traced the term "blog" back to 1997 using the Google Archives. "Weblog" had the earlier meaning of a log or record generated by software of web users accessing a site and was chosen by Jorn Barger in December of that year as the name for his new page of interesting links to news stories, probably because weblog as an automatically generated record of users suggested the possibility of a manually generated list of links (see also Blood, 2002). The word was shortened to blog in 1999 after the suggestion of its pronunciation as "we-blog" (Blood, 2002).

Rebecca Mead put blogs into the mainstream in November 2000 with a well-written story in the *New Yorker* about a leading blog site (Mead, 2002). According to a history of blogging, as recently as 1999 there were fewer than 25 blog sites, while by 2002 one leading portal listed over three thousand, and the MIT Media Lab Blogdex Project currently tracks over 14,000, and the most popular blogging site, blogger.com, boasts some 150,000 users (Blood, 2002; Sullivan, 2002). Some blogs are organized around hobby or interest groups, while others focus on political or protest events, for example daily blogs from the Davos WEF meeting in New York City during January, 2002. These online daily updates shade into the longer-standing reports distributed via email and discussion lists, and connect back to the days of personal fanzines (perzines), flyers, bullhorns and eighteenth-century pamphleteering. Indeed blogging can be interpreted as the modern digital equivalent of the memoir and diary (literally "daybook").

Why not simply write these musings into a text file or offline diary? Most bloggers would find this a strange idea. One is writing not as an academic exercise but to be in a community. This community is not "the mass", but rather a set of people (constantly changing although it is) that pay attention to each other's blogs and link them. There are distinct sub-groups defined solely around the fact that they like each other's writing. Feedback on the day, so that feedback itself becomes part of one's experience, is critical.

Blogging is an activity which both takes place in and produces community. Bloggers link to each other, comment on each other's site, mention each other in their blogs, create "fansigns" (buttons or cool graphics mentioning the site's name, or webcam pictures with the site's name inscribed somewhere on the body) thus creating friendships and mutual support. Of course it's not all sweetness and light. Where communities intersect there can often be disputes or internal debates about how best to handle an issue or someone's opinion (it's also notable that if a blogger is criticized strongly in the comments section, the criticizer is often well outnumbered by supportive comments). There are also portals to blogs, and categorizations of blog types, rankings, conventions (Weblogcon 2002) and so on, in other words blogging is now a fairly mature and increasingly commercialized practice (although with much resistance to the latter). Blogging communities in fact, although now commonly hosted on specialist blog-server sites and using standardized web page software, are themselves free. The blogger employs some interesting strategies to pick up a few dollars, for example by allowing porno websites to have links on their site, or providing "wishlists" from which people can buy gifts for the blogger. Blogging constitutes an especially fascinating practice of the self "in" the world-as-community through self-writing.

Clearly care of the self is not exhaustively constituted as personal confession and indeed confession can refer to a number of interrelated practices. While the outcome of "classical" confession – the "queen of proofs" as it was called in medieval times – is to produce authentic discourse or the truth about oneself, self-writing such as blogging has no such target. Rather, they are part of the process of a "life emerging" as one diarist puts it.[3]

ON CONFESSION AND CYBERSPACE

> Sex is boring . . .
>
> Confession of Foucault (1997b, p. 253)

I have argued that much of our understanding and interpretation of cyberspace has proceeded through the twin concepts of confession and authenticity. To counterattack these problematics, in the last chapter I used a Heideggerian notion of authenticity that does not constitute authenticity as identity. We now need to see why confession requires us to accept that it is possible to produce the truth about ourselves or our world and why this is a narrower and more troubling approach.

To confess is to be in a position of (1) being authenticated as who you say you are (real/false) and (2) being placed in a discourse of normalization. As Rose has noted:

> Confession was the diagram of a particular form of power . . . the truthful rendering into speech of who one is and what one does – to one's parents, one's teachers, one's doctor, one's lover – was both identifying, in that it constructed a self in terms of a certain norm of identity, and subjectifying, in that one became a subject at the price of entering into a certain game of authority.
>
> (1996, p. 96)

To see whether we can change the rules of this game of normalization and authority it is first necessary to understand confession a little more deeply.

In the traditional juridico–religious model (e.g., in the Catholic church) one confesses to an authority with true penitence (sorrow and admission of guilt) in order to remove the sins which have come to lodge with one's soul. The removal of these sins through the production of truth about oneself (constituted as the examination of one's conscience and confession itself) enables one's true, forgiven self to emerge and shine forth. The problem with this model of self-production is that it is heavily enmeshed in procedures of normalization. Even when confession becomes "scientized" and medicalized and the notion of sin drops out (for example in psychiatry or the legal confession), by speaking one's mind frankly, our desires or acts such as longings for another person, are part of a process of the "normal" and the "deviant". Your confessions are assessed in terms of what is normal and what is not so that you may receive the proper, authentic treatment or punishment. The treatment is truly and authentically yours, and you are authentically produced between the poles of normal or deviant. Are you depressed or not? Are you manic-depressive? Do you have unhealthy obsessions? Did you practice normal or deviant sexual acts? With each diagnosis you become a juridico-medical case within a scheme of normalization.

It is clear that the West is a highly confessionalized society. We like to make confessions and to listen or watch them on TV, in the news or at church. The chat shows on US daytime TV such as *Oprah, Sally Jessy*

Raphael, Jenny Jones, Ricky Lake, and *Queen Latifah* provide ample opportunities in which to confess one's transgressions or secret yearnings.

There are many types of confession from the venial to the mortal, but there cannot be any doubt that as a society we have created more space and opportunity to perform confession than at any other time in history. The modern multimedia industry (cable, TV, radio) helps to achieve this, and we find it reflected in the law (the interview room), schools (the principal's office), the analyst's couch, water-cooler conversations, and our religion. Yet there is, surprisingly, no field of "confession studies" which takes the practice of confession as its subject, except in the narrow instance of historico-literary studies (i.e., "confession in the novels of . . ."). This is even more surprising given the vast literature on the "self" in general, including but not limited to psychiatry, psychology, philosophy, care of, the formation of the self, and on. What motivates someone to speak their mind? Are confessions voluntary, and if not, what makes them involuntary? Are they always involuntary for the same reasons historically or spatially? Are there false confessions and, if so, what procedures have been developed to identify them? What are the interplays of truth and lie, guilt and innocence? Finally, how does the performance of confession create a community – is confession the "ground" (upon which, or grounds for) a community roots itself?

These questions, and no doubt many others, would form the target and object of some future confession studies program and cannot all be addressed here. But as Brooks argued recently, confession

> is deeply imbricated with our sense of the self, its interiority, its capacity for introspection, self-knowledge, self-evaluation. If we, as a culture, are going to attach such importance to it, see it as so definitional of the person, we need to scrutinize the conditions of its production and the source of its articulation.
>
> (Brooks, 2000, p. 171)

Instead of practices which produce the truth concerning oneself in a field of normalization (traditional or juridico-religious confession) cyberspace can be the ground for self-writing in communities which effect an ethos of care of the self.

Let us consider some of the ways in which confession gets implemented in cyberspace due to the requirements for authentication. How does the need for authentication call forward confession? There are two broad mechanisms, one in which we exploit the many increased opportunities for confession, and one in which we "renounce" ourselves.

First, cyberspace offers boundless opportunities for the *extension* of confession from its long history in Western society. Our society is a highly confessionalized one, although no doubt the particular form and rationale of confession has changed historically, lapsed in and out of public consciousness, and wavered between secular and juridico-religious manifestations. In cyberspace there are explicit places and spaces online in which to perform juridico-religious confession, discuss confession, or view confessions (e.g., see notproud.com, stthomasapostle.org/Pastoral/askapriest.htm, and dailyconfession.com, as well as the many outreach websites of local churches). Yet these are relatively specialized and niche spaces. More significant, as we have already seen, are the procedures for authenticating one's true identity with passwords, encrypted network protocols, Acceptable Use Policies designed to prevent password sharing, Internet Protocol (IP) address checking, email-header tracing (to eliminate and identify the origins of emails), individualized logins, and so on. These procedures are all targeted at producing and authenticating the true identity (the inner self) of the prospective user and to differentiate one user truthfully from another. That is, they are confessional procedures in the juridico-religious sense which produce the truth about the self.

Second, some commentators (for example, Haraway, 1991; Turkle, 1995) have suggested that cyberspace is *necessarily* confessional in that we always already "renounce" our bodies entirely in order to free the inner true self. On this argument the body is conceptualized as being left behind or discarded in the virtual, as it was for some early Christians in their practice of publicly showing yourself (an *exomologesis* or self-renunciation): "you will become the subject of the manifestation of truth when and only when you disappear or you destroy yourself as a real body" (Foucault, 1999, p. 179). For example, a disabled person may navigate the highways and chatrooms of the Internet, or any able-bodied person can go beyond material distances to meet people far away. Or, an abused woman may find safe spaces online, free of threats to her bodily person. In this view, cyberspace allows people to move beyond the binary hierarchies of gender relations and the bodily to be their true selves. Such was the viewpoint articulated by Rheingold in his influential book on virtual communities (1993): "people whose physical handicaps make it difficult to form new friendships find that virtual communities treat them as they always wanted to be treated – as thinkers and transmitters of ideas and feeling beings, not carnal vessels with a certain appearance" (p. 26). For Rheingold, the self is essentially the Cartesian mind within a body, and cyberspace frees up this mind and renounces the body. There is the essential (inner) subject and true self, and the (external) world. On this view, by definition the digital is precisely where the body does not enter, but where your true inner self can shine forth. The digital is an entirely separate realm, opposed to the physical. We may enter cyberspace, but we will have to leave behind, or renounce, our body, which therefore allows our identity to become multiplicitous.

In their discussion of the ongoing construction of the self in cyberspace, Dodge and Kitchin (2001) note that it while this might accord with "current psychoanalytical and postmodern theories of identity" (p. 24) we should be careful about too uncritically adopting this view outright. As they note, it has the unfortunate result of promoting cyberspace as a separate realm, an objectivized space of Paul Virilio's "real time" and end of space. While there is some attraction to the idea of slippery identities that are constructed at will, I have also emphasized that as being-in-the-world we exercise this ongoing project of ourselves most authentically when we take up our own possibilities. Ontologically, we are always already "in" a world not of our own making: we have to begin from there and not some unlimited pool of possible identities.

Space (including cyberspace) is not a realm where distant selves form virtually proximate communities through communication (the truthful confession), but a world where humans are with each other and *into which* various technologies such as telephones or chatrooms are introduced (Coyne, 1998). Thus a shared world exists prior to technology. Digital technologies may interplay and reconfigure the primary shared world, but do not originate or form it (in fact they are called up because they can be useful to promote our being with each other and the world). Even in cyberspace we are still being-in-the-world. Thus the idea of the abandoned body, intriguing though it is, is difficult to sustain. It again constitutes cyberspace as a separate heterotopian space in opposition to the physical, rather than as part of our world in general. Authentication procedures act as gatekeepers to this world, checking and letting in only those with verifiable identities. The implications for the spatial politics of cyberspace are significant because at issue is the way the self is *placed* in the world and the manner in which we have an authentic understanding of that world. Authentic place is only produced when the spatial representations of (cyber)space, that is our maps, have confessed the truth of the landscape.

Thus, the authentication–confession nexus (in either its relative or essentialist versions) produces a certain kind of truth about subjects and place. It is one in which subjects are discriminable individuals with identifiable selves, who dwell in physical space, and who produce the truth about themselves in order to enter the separate domain of cyberspace, and that this truth is interpreted in a framework of normalization (authentic–not authentic). Far from the postmodernist celebration of cyberspace enabling fractured selves with free-floating identities then (e.g., Turkle,

1995) the authentication–confession nexus of cyberspace today produces particular normalized subjects, and constitutes cyberspace itself in opposition to physical space. We need to challenge both of these interpretations, not in order to refracture the self, but to practice the self. Thus cyberspace is bodily, in that embodiment is part of its essence because it is not a separate domain of the virtual but part of the self in the world. But its interpretation as confession of inner truths in response to the authentication crisis has eclipsed this view, and set up cyberspace as a separate space opposed to the physical.

What are the characteristics of power in confession? Confession takes place in the presence of a partner who is not just the interlocutor, but an authority. This figure regulates the confession with two interesting effects; first it casts confession as all the more valuable because it is difficult. Second, confession produces deep differences within the confessing subject – he is unburdened, purified, relieved. Although the listener is the dominant partner, the production of truth occurs in the very process itself and the effects take place on those who speak, not in those who receive it. Skills are not communicated with an effect on those who hear them, but the effect is upon the speaker.

Foucault had earlier discussed how confessions were performed under duress, or even torture, in his book *Discipline and Punish* (Foucault, 1977). Around the time of the twelfth and thirteenth centuries in Europe confession became the dominant mode of finding the truth in judicial cases (not coincidentally at the same time as the 4th Lateran Council codified confession in religious practice – see the Catholic Encyclopedia, 1913). Foucault notes three essential features which are required for punishment to be torture:

1. It must produce a certain degree of pain. This pain should be quantifiable and calculated.
2. Torture must be regulated and in proportion. It is in proportion to the gravity of the crime, the standing of the criminal, and the rank of the victim.
3. Torture is part of a ritual. First, it must mark the body of the criminal, or through public humiliation, to brand them with infamy. Second, torture takes place in public as a grand spectacle, almost of excess. "The fact that the guilty man should moan and cry out under the blows is not a shameful side-effect, it is the very ceremonial of justice being expressed in all its force" (Foucault, 1977, p. 34).

Why was torture used for so long in the production of truth in criminal cases (say until the criminal reforms of the late eighteenth century in Europe)? Again, we can look to the circulation of power to explain this.

Torture "revealed truth and showed the operation of power. It assured
. . . the procedure of the investigation on the operation of the confession
. . . it also made the body of the condemned man the place where the ven-
geance of the sovereign was applied, the anchoring point for a manifestation of power" (Foucault, 1977, p. 55).

The effects of the confession were given a scientific rubric; the confession was not just a documentation of sins, but was adjudged as either normal
or pathological, and, if the latter, could be used within therapeutic treat-
ments: the truth could heal (Foucault, 1978, p. 67). It can be seen how this
would bring into play the expert, who could act as gatekeeper in a process
of normalization. In the case of penal torture, this gave rise to the person of
the inquisitor, in the case of psychiatric medicine that of the psychiatrist.

As we shall see, many of these suggestions have applications in other
fields, including cartography or cyberspace, because what Foucault is
talking about is the circulation of information and systems of representation, as well as the production of truth and power relations in the construction of the subject. Another way to think about this is as the shifting
and gapping of the line between public and private (confession makes
private things public).

CONFESSION THROUGHOUT CYBERSPACE

As for cyberspace, the distribution of confessional possibilities has never
been greater. We may begin with sites on the Internet which emerged to
explicitly encourage confession. There are several of these, such as not-
proud.com and dailyconfession.com. For example, notproud.com, which
went live in Fall 2000, has by its own count attracted over one million page
views (or accesses). Notproud.com allows visitors to anonymously leave a
confession under one of the "seven deadly sins" (pride, envy, sloth, lust,
etc.) and for these confessions to be read. No doubt some of them are jokes
or false confessions (as sometimes admitted later in other confessions) but
the simple collection of human life in this site makes compelling reading
for the sheer audacity of what people will admit. For example:

> I was at the grocery store today. A woman in front of me took the
> last box of little debbie brownies that i wanted. While her back
> was turned, I spit on her baby and took something out of her cart.
> It turned out to be olive oil – which I am allergic to – a pyrrhic
> victory of sorts.

True or not?

Quite a few entries provoke an outburst of frank laughter. Of course this is only one site, even if it is an epitome of the confessional archive, and different in kind only because the confessions are not kept confidential (although anonymity is guaranteed).

Confession need not be entered into "willingly" (as in the case, at least in theory, of confession to a priest). There are many strategies and practices which seek to elicit the most exacting of confessions in order to be authenticated. I do not mean actual torture, but where one "must" confess to an authority in order to take the most simple action. Barbara Ehrenreich's book *Nickel and Dimed* (2001) on the conditions of the working poor reveals how job applicants are subjected to a whole battery of questions and physical tests which have nothing directly to do with the job. Instead, the

> real function of these tests, I decide, is to convey information not to the employer but to the potential employee, and the information being conveyed is always: You will have no secrets from us. We don't just want your muscles and that portion of your brain that is directly connected to them, we want your *innermost self.*
>
> (Ehrenreich, 2001, p. 59, emphasis added)

Urine and drug tests are designed "to involuntarily reveal the innermost self" of the applicant rather than assess their job skills. In an interesting reverse, one of the first signs that you have lost your job at a dot–com is that the computer will no longer let you log on – you have been de-authenticated. Davidson tells the story of a worker for an Atlanta dot-com which was going under:

> [W]hen Haraway showed up for work early on June 22 [2001] and was denied access to the company's computer network, he knew what it meant. A couple of co-workers fired up their PCs only to be similarly blocked. It's fittingly symbolic that being cut off from e-mail and a computer often is the first tangible sign that you no longer work for a technology company. Essentially, the machine is telling you're fired.
>
> (Davidson, 2001, p. 30)

A special website "F*ckedcompany.com" tracks rumors of these dot-com layoffs. Yet, as we have seen, these sites where one is implored or enjoined to confess are only isolated instances of the confessional in cyberspace. Contrary to the Vatican, cyberspace is marked through with confession.

As discursive and confessional communities develop in cyberspace we might begin to wonder about the spatial politics of these communities. How do their membership criteria and social networks promote a sense of belonging-in-place or, on the other hand, a sense of exclusion and elitism? Is it more rewarding to join a pre-existing community or to create one's own and attract other members through the honesty and creativity of your writing? In other words how does will to authentication intersect with power relations and the production of a sense of place? In the remainder of the chapter I wish to take up a particular aspect of this confessional authentication, namely that of resistance, in order to point toward some possible techniques that permit or encourage the practice of the self in relation to power. These are the related topics of self-writing and frank-speaking – or as the Greeks called it, *parrhesia*.

RESISTANCE: BLOGGING AS SELF-WRITING

"I still have a world of me-ness to fulfill"
                              Cocteau Twins, "Half-Gifts"

What is "resistance" in cyberspace? As a technology of the self, practices such as blogging are a form of resistance to normalization because they are where one works on oneself in a process of becoming. It hardly seems likely that one can work on oneself as a project if one is completely dominated by power (Weberman, 2000). Usually, the object of resistance is to problematize something which was not previously seen as a problem, in order to allow both new sets of knowledge but also new ways of being, new practices. But these ways of being must already include the possibility of problematization. So problematization is an essential part of resistance. Being must already be problematizable.

Commentators have often focused on those parts of Foucault that offer hints on resistance. These include his writings on "transgression", the "insurrection of subjugated knowledges" at the local level (Foucault, 1980b, p. 81), his involvement over the Iranian revolution, whether liberation is possible and if it is sufficient, and no doubt many others. In his last period of work, which, I assume, is not to be taken as the final word, Foucault at least makes it explicit that resistance, indeed the production of the truth of oneself, was an ongoing project.[4]

We have already suggested that blogging is best understood as a technique of the self being-in-the-world, of being in community with shared meanings. Now that we have grasped the problematic of the confessional I return to blogging as a deliberate strategy of resistance against the normalized, confessional conception of the self.

What does a blog look like? Each blog is different, but commonly features a daily entry, link to something interesting, a rant, or simply, just chatter and musings. Most blogging sites let readers comment on the entry. Some are like bulletin boards, and some have so many responses they are like chat rooms. Often these responses are on another part of the site, which allows the original blogger to maintain priority for their writing (it doesn't get buried under the responses). There's no point trying to define the content because it can and does include nearly everything. Some blogs are themed, e.g., war blogs or Andrew Sullivan's political blogs (Sullivan is former editor of *The New Republic*). Many thousands more take advantage of this small revolution in self-publishing to simply take themselves as the theme – not necessarily in a selfish way (many blogs talk about how the author relates to other people or events), but as part of living one's life. The point of the blog is not the content: it is the form, or process itself: self-writing.

But other blogs are more akin to diaries and online journals, and since they are updated on a daily basis they are quite literally quotidian. As one site explains:

> This is a personal website, meaning that we are not selling a
> product or advertising a cause. It is simply a journal of sporadic
> thoughts of two young people. We aren't "living double lives". We
> are not "Internet celebrities". We are *regular* people that write
> *regular* stuff on the internet. Meaning that, no, we don't like being
> stalked by you or hounded by paparazzi. I'm keeping a little part
> of the web to collect my thoughts.

For these authors the blog is not a separate part of their lives, as if they had a private and public life. The blog is an ordinary, everyday aspect of who they are. Another popular blogger wrote in his blog in 1999:

> You see, CamWorld [his blogsite] is about me. It's about who I am,
> what I know, and what I think. And it's about my place in the New
> Media society. CamWorld is a peek into the subconsciousness that
> makes me tick. It's not about finding the most links the fastest,
> automated archiving, or searchable personal web sites. It's about
> educating those who have come to know me about what I feel is
> important in the increasingly complex world we live in, both
> online and off. CamWorld is an experiment in self-expression.
>
> (Barrett, 1999)

For Barrett blogging is a practice of self-expression, in public, in which he can investigate his "place in the world". In other words, blogging is a practice of the care of the self through techniques of self-writing. These techniques produce authenticity of the subject. Many bloggers get pleasure from blogging, it's a voluntary way of making your mark (individualizing) and in the supportive feedback that bloggers receive is a sense of validation (or authentication if you prefer). Blogs are deliberate, highly personal writings. Ethnographic and socio-political analysis of blogging (perhaps employing some of the emerging blog meta-indices) would say a lot about the context and content of these discourses, as well as the power relations in which they are embedded. Some blogs are deliberately oppositional, designed to promote political discourse which bypasses the traditional media outlets in a more democratic manner (akin, perhaps, to Habermas's idea of communicative action). As one journalist-blogger wrote on the question of why blog:

> Because as much as I love newspapers, magazines and broadcast media, I'm not happy to be a passive receptacle, a recipient of packaged product. I want to be a participant in the news, a party to social discourse, a part of the conversation.
>
> (Lasica, 2001)

I want to be a participant in the news. Thus participation creates the self as a more fully developed person, one who is not passive toward him or herself, but part of the conversation of human discourse. They want to develop themselves rather than expose a previously hidden truth about their innermost self. In this aspect they are far more similar to the Greek Stoic writers such as Seneca and Epictetus than the later Christian confessors, marked as the latter were by confession to an authority for the purpose of admitting sins or unlocking desires. As for Seneca, today's blogger is playing in a truth-game whose objective "is to make of the individual *a place where truth can appear* and act as a real force through the presence of memory" (Foucault, 1999, p. 168, emphasis added). This kind of practice Foucault called the gnomic, in the sense of a coincidence of the will and truth; the gnomic self is one where there is a will to truth, or better, of truth (not just on the way to truth, but actually having truth in conjunction with the willing self). We can understand this in distinction to the Christian practice of confession as the problematic where "the self has, on the contrary, not to be discovered but to be constituted" (Foucault, 1999, p. 168) and it is constituted by blogging as a practice of pleasurable self-writing. No doubt future studies will show further sociological elements of this self-writing, for example how surveillance and data collection about ideas and persons may be collected, but that is all I shall say about blogging for the time being.

RESISTANCE AS *PARRHESIA*

Parrhesia is a Greek word that means frank or fearless speech, especially when spoken against those who are more powerful. It is first used during the fifth century BC in the tragedies of Euripides (especially *Ion* and *Orestes*). One speaks freely because one has a concern for the truth, even at personal risk to oneself. It is a self-disclosure, but unlike confession does not have the purpose of excavating the deep self in order to yield it up to God or eradicate the self. *Parrhesia* is used to show that you can give an account of yourself in the Socratic sense (Foucault, 2001), i.e., that there is a relation between the way you live your life and rational discourse (the *bios* and *logos*). It is a rational accounting of someone's life.

Foucault's lectures in 1983 and 1984 investigated what it meant to tell the truth, even if that incurred a personal cost. Foucault defines *parrhesia* as having the following properties. First, it means frankness; speaking everything on your mind and not holding back (Foucault's etymology is *pan* "all" and *rhema* "that which is said" Foucault, 2001, p. 12). Second, parrhesia is the practice of telling what is true; "the parrhesiastes says what *is* true because he *knows* that it is true; and he *knows* that it is true because it *is* really true" (p. 14). Here Foucault makes a distinction between Cartesian evidence-gathering in order to determine what is true because one is doubtful about what one believes (data collection and analysis) and the Greek notion that truth obtained to those with certain moral qualities. Third, it is necessary that there be risk or danger in telling the truth. Foucault gives the examples of a philosopher addressing himself to a tyrant, such as Plato and Dionysus in Syracuse, or of a citizen criticizing the majority. The risk may be personal danger, loss of friendship, or popularity. Therefore one does not speak the truth for truth's sake, but for purposes of criticism, either to another or to oneself; *parrhesia* acts from below, among those with less power, and is directed "above" at those with more power (p. 18). Finally, the practice of *parrhesia* is one of duty and moral obligation. It is a freely made choice and cannot be compelled (a forced confession is not parrhesiastic). Or again, a criminal forced by the authorities to confess his crime is not using *parrhesia*. *Parrhesia* "is thus related to freedom and to duty" (p. 19).

Several times Foucault is careful to state that *parrhesia* is quite different from confession. Foucault argues "because we are inclined to read such texts through the glasses of our Christian culture, however, we might interpret this description of the Socratic game as a practice where the one who is being led by Socrates' discourse must give an autobiographical account of his life, or a confession of his faults. But such an interpretation would miss the real meaning of the text" (Foucault, 2001, p. 96).

But *parrhesia* is also different from rhetoric and this opposition lasted for several centuries. Foucault's timeframe in these lectures is the fifth century BC through the fifth century AD, and *parrhesia* is separate and preferable to rhetoric until at least the second century AD (p. 21).

On the other hand, it was essential for politics, for a true democracy. One had to be able to make public speeches in the agora during Plato's time. Later, in Hellenic times, *parrhesia* becomes more individualized, and characterizes the relationship between a sovereign and his advisors. So there are several possible types of *parrhesia*: the political, and the monarchic. Foucault will also add the Socratic, where, he says, it is regarded as an art of life (*technē tou biou*).

So far, these characteristics of *parrhesia* have concerned the obligations and risks of the one who uses *parrhesia*. But perhaps equally interesting, especially how we might think through parrhesiastic cyberspace, is what Foucault calls the "parrhesiastic contract". He illustrates this with reference to another of Euripides' plays *The Bacchae* (407–406 BC). Here the king's servant, a herdsman, asks the king whether he may speak freely because he has bad tidings to bring and he does not want the king to get mad at him. The king is not a tyrant, but a wise king, who understands the parrhesiastic game and offers his servant the parrhesiastic contract. Therefore there is in this contract a kind of complementarity, because the one with power lacks the truth, while the one with the truth lacks power. Hence the contract; the servant gets some measure of protection (though not a complete one), while the sovereign, who retains his power, also gets the truth. But it is a sign of the strong sovereign that he can take criticism in public.

What are the implications of a parrhesiastic contract for cyberspace? There seem to be both desirable and less desirable ones. On the one hand, if the "powers" of cyberspace (the sovereigns) are wise and cooperative, they will allow people to speak and publish freely, to let all speak, not just an elite, and to govern with a light hand. On the other hand, it reinforces the powers that be, encourages the stability of power, and legitimizes the ruler–ruled relationship. In other words, you could say that *parrhesia* is complicit with power, rather than resistant to it. If we recall that in Foucault, power is never power as such, but always relations of power, then the particular relation here (in the contract) is typical of power; there is a relationship between those that control and the controlled, it is not just a question of domination. And furthermore, other power relations are possible, which Foucault discusses later, under the topic of Cynics and Diogenes, which could be called resistances or insurrections at the specific, local level. In cyberspace therefore, we understand that the ones without authority operate under certain conditions, such as the fact that email is not private, or that there are racial differences in access to computers, but they are accepted because they afford a certain advantage as well.

There are two examples of *parrhesia* I would like to highlight. I want to lay these out as a way of creating or sustaining a *space of resistance* for cyberspace.

There is one well known example and one lesser known. The first concerns that of Diogenes and the Cynic philosophers, who were prevalent from the first century BC to the fourth century AD, or some 500 years. The word cynical now is almost entirely negative, but for the Cynics themselves it meant more a way of life than a philosophy. This way of life was meant to embody or constitute the truth, so that there should be as close a match as possible between word [*logos*] and deed [*biou*], or between theory and practice. The Cynic tradition was deeply embedded in Greek thinking and it has relations to the Stoic and Socratic traditions as well as to Christianity (there is a modern Jesus as Cynic hypothesis). The way of life they adopted was one of calling into question established modes and hierarchies (we might say of critical thinking or problematizing). This was achieved through public preaching, satires, scandalous behaviors, and *parrhesia*. The Cynics were quite independent and prized freedom [*eleutheria*] and self-sufficiency [*autarkeia*, literally self-rule]. There was a wide libertarian streak, in that they saw all dependency on culture, institutions, or society to be obstacles to a natural life-style (the way of the dog, or *kuōn*). But it was not an isolationism – they wanted to call things into question, not just do away with them altogether in any form.

One of the ways they did this is by a sort of provocative discourse. Dio Chrysostom, a quasi-Cynic of the first and second centuries AD depicts Diogenes as moving to the very limit of the parrhesiastic contract. For example, in his dialog with Alexander the Great, Diogenes starts by calling him a bastard and challenging whether he has the true qualities of a king! Naturally Alexander does not like hearing this, and no doubt as king, could kill Diogenes for saying so. Therefore Diogenes takes a great risk in speaking as he does. But Diogenes does not quite cross the limit of the contract, and each time Alexander gets angry Diogenes comes back with a couple of techniques to rope Alexander back into the game; he acknowledges that Alexander could kill him and he offers him a choice; kill him, the only one who can tell Alexander the truth, or let him live and learn the truth. This re-establishes a new parrhesiastic contract, because Alexander is impressed with Diogenes' courageousness and lets him live. Therefore the line has now been moved, and Diogenes is freer to speak. It's like Diogenes has now negotiated that he will not be killed. A second strategy is that Diogenes uses charm in a subtle way so that his interlocutor believes he has heard something complimentary. Diogenes uses this charm, not to flatter, but to convey another truth, or to calm Alexander for another round of aggressive exchanges. All this can be found in Foucault's discussion of *parrhesia* and the public life (especially the fifth lecture, Foucault, 2001).

It's clear that the strategy is one model of resistance that can be used for cyberspace and more generally too. In this model one challenges established norms right up to the level of the contract. Then, when this limit is approached, one ups the ante so that the ground moves, or the condition of possibilities is shifted or enlarged.[5] As Sloterdijk puts it: "[w]ith Diogenes, the resistance against the rigged game of "discourse" begins in European philosophy" (1987, p. 102).

This is different from ordinary resistance. For example, ordinary resistance, which is also very important, might consist of constantly challenging the validity of laws which restrict free speech in cyberspace. Organizations such as the ACLU, the EFF, CPSR (Computer Professionals for Social Responsibility), and the American Library Association have done this by funding legal challenges (American Library Association, 2001). But these are largely reactive to particular instances of power. In the Cynic model the grounds of knowledge, the terms of the debate, are changed as well as the practices. I think there is a prime example in the case of debates over the Internet and decency. Organizations such as the Klaas Foundation and the US Department of Justice construct a set of discourse about "safety", "dangerousness", "perversity", and so on in a very unproblematic way. Given these terms a set of remedies (Internet restrictions, spatial surveillance and crime-mapping, sex offender lists posted on the Internet) appears very "naturally". Even to challenge the appropriateness of these terms can be risky. So what is needed is to shift the line of the "contract" so that these terms can be problematized or thought about critically, that is to reveal the governmental rationalities behind them that give them their status as truth.

My second example is a kind of modern equivalent of the Cynic and it is that of the science fiction writer Philip K. Dick. Dick's name is very well known in SF, but little read outside it. In fact he's better known for the films that have been made of his novels, such as *Blade Runner*, *Total Recall*, *Screamers*, and *Minority Report*. Many of today's better-known authors, such as William Gibson, Bruce Sterling, Neal Stephenson, and other cyberpunks, were influenced by Dick's work (he died in 1982). Dick did not self-identify as a parrhesiast, unlike the Greeks. But Dick does have a moral duty to seek and tell the truth, he often "spoke truth to power" which put him in a dangerous personal situation. He also wrote an extremely lengthy "Exegesis" which was not meant for publication, but was a place where year after year, he worked on his thoughts and tried to understand himself. It could certainly be described as a frank saying-all. In fact he sometimes saw this as the science fiction "style" itself:

> The reason why I'm a freelance writer and living in poverty is
> (and I'm admitting this for the first time) that I am terrified of
> Authority Figures like bosses and cops and teachers . . . but at the
> same time I resent it – the authority and my own fear – so I rebel.
> And writing sf is a way to rebel. I rebelled against ROTC at U.C.
> Berkeley and got expelled . . . later on I was to oppose the
> Vietnam War and get my files blown open and my papers gone
> through and stolen.
>
> (Dick, 1980, p. xiii)

Very early in his career (c.1952–53 while at Berkeley) he and his second wife Kleo were approached by the FBI and asked to inform on enemy agents at the University of Mexico. Dick was conflicted about cooperating with the authorities and castigated himself in the "Exegesis" for his cowardice (Sutin, 1991, p. 217), while at other points congratulating himself on his staying power as a counter-culture "quasi-Marxist" (Sutin, 1991, pp. 174–5). It is certainly true that the bulk of Dick's life and work was dedicated to anti-authoritarian concerns, both by his nature and by geography (Berkeley, Southern California, etc.) and by chronology (Nixon, Watergate, McCarthyism). In this sense Dick put the "punk" in cyberpunk.

Dick opposed himself to authority, and took up a marginalized literature where he could speak for the common man. His stories often feature ordinary people struggling to survive in hostile environments. For example, in his novel *The Three Stigmata of Palmer Eldritch* colonists on Mars must battle against the barren landscape to make things grow, and turn to drugs which create escapist fantasies.

Dick frames these concerns around the question of what it means to be human. Dick begins by noting that the distinction between the human and the artificial construct is blurring. What, then, is the essence of humanity? Is it free will – no, for much of what we feel and believe may be illusory. Surface appearance or biology is also no guide, for a person might be biologically human but not "authentically human" (Dick, in Sutin, 1995, p. 187) because they exploit others. Such exploitation is typical of governments which control their populace, of over-commercialization and media which acts to conform individuals, of laws which restrict the young from voting, and of technological devices which surveille and direct.

Conversely, Dick takes delight in populating his fiction with mechanisms which render assistance, such as Dr Smile, a friendly psychiatrist-in-a bag, or the seemingly repellant, which may possess great *caritas*.

Dick concludes therefore that "humanity" refers "not to origin . . . but to a *way of being* in the world" (Dick, in Sutin, 1995, p. 212, emphasis added). This way of being involves change and becoming as a result of experience – unlike the autonomic mechanism or animal which cannot become: humans can be "*changed*, altered, what it did and hence what it was; it *became*" (Dick, in Sutin, 1995, p. 203). Not only is this akin to the Foucauldian sense of ethics as a concern for developing yourself, but it is reminiscent of Heidegger's existential analysis of being (as far as is known, Dick did not read Heidegger or Foucault).

Dick's positive response to power was to encourage resistance or what he called "balking" (Dick, in Sutin, 1995, p. 278). In terms remarkably reminiscent of Foucault's discussion of *parrhesia*, Dick wrote in 1978 that:

> The authentic human being is one of us who instinctively knows
> what he should not do, and, in addition, he will balk at doing it.
> He will refuse to do it, even if this brings down dread
> consequences to him and to those whom he loves. This, to me, is
> the ultimately heroic trait of ordinary people; they say *no* to the
> tyrant and they calmly take the consequences of this resistance.
>
> (pp. 278–9)

Dick most fully developed the balking principle of the authentic human in a 1977 interview (Dick, 1987). Imagine, he says, the worst possible social situation one can be in, that of Nazi Germany (Dick had done much work on this for his award-winning 1963 novel *The Man in the High Castle*). The German civil police were given lists of Jewish people and brought them in. Clearly, the municipal police (unlike the Gestapo) had no idea what would happen to the Jews who were brought in. But, says Dick, they did not stop for one moment to question why – where were they going? for what purpose? who says? Dick equates complicity with the authorities with a mechanical, androidal mentality, which unfortunately was also present among the victims, who went along passively. On the other hand, Dick says:

> The idea occurred to me that the human element would come in
> where the policeman would say, "Wait a minute. I want to know
> what you're going to *do* with these people". I got this idea of
> *balking*; this very vivid picture of a human being suddenly
> stopping in his tracks . . . saying "No, I won't do this. I won't go
> get these people".
>
> (Dick, 1987, pp. 46–7)

Prefiguring Kenneally's *Schindler's List*, he imagines an everyday worker, "Some secretary, for instance, who has to type up the list. All of a sudden she could say to herself, "No, I'm going to leave a name out . . . or I'm going to leave two names out . . . or I'm going to give the address wrong," or something like that". The effect of balking, of resistance to power from the ground up, creates possibilities of change. "All that really would have been required", says Dick, "would have been some balking along the line by people here and there, and the structure would have begun to become unglued" (Dick, 1987, p. 47). For Foucault, as with Dick, resistance worked from the ground up. One does not have power or use power; it exists in a dialectic with resistance. *Parrhesia*, as expressed by Dick in his balking principle, is a historical example of this local resistance.

Dick further explored this idea of balking and resistance to authority in his short story "The Exit Door Leads In" which was written around this time (1979) and published in a college-oriented magazine (Dick, 1992). In it, the major character Bob Bibleman wins the chance to attend a military college where he is abused and told to obey commands. While there, he stumbles on an engineering plan which could provide unlimited quantities of cheap energy, but, out of loyalty, decides to turn the plans over to the authorities which he knows will suppress them. He thinks he has done the right thing. But in a final interview, he is told that he has actually failed the test, which was really to see if he could act as an individual by *resisting authority* despite the possible personal consequences: the plans were bogus. In speaking to the next generation, Dick was clearly suggesting the life of the parrhesiast.

In Dick's 1972 essay he once again sees this parrhesiastic truth-telling in the youthful hacker mentality, who have grown resistant to the importuning of the media and are too rebellious to listen to the authorities. Dick cites with approval the case of phone phreaks such as "Captain Crunch" who bypassed the telephone company's system in order to make free calls. Dick as a science fiction writer finds it ironic that it is technology which has made much of this control possible, and warns that:

> If, as it seems we are, in the process of becoming a totalitarian society in which the state apparatus is all-powerful, the ethics most important for the survival of the true, human individual would be: Cheat, lie, evade, fake it, be elsewhere, forge documents, build improved electronic gadgets in your garage that'll outwit the gadgets used by the authorities.
>
> (Dick, in Sutin, 1995, p. 195)

Dick wrote this at the end of the Vietnam War, and he speculates on what would happen if its technology such as sniperscopes, passive infrared scanners, and surveillance devices fell into the hands of the people being monitored and were turned against those who developed them. There is a certain naïveté here (think of the Unabomber, McVeigh, or the September 11 terrorists) but also a certain attraction to the idea of the inmates running the asylum, or an inverting of the panopticon.

Sure enough, these parrhesiastic ideas subjected Dick to criticism and the attention of the authorities (Dick was visited by the FBI in the 1950s and accrued a minor file for his 1968 participation in a left-wing magazine). There was a sense at the time of the oppressive police state (what Dick called the Black Iron Prison), which is certainly reflected in his writings and his letters sometimes to the point of paranoia. Thus humanity for Dick can be defined as those who exhibit an ethics of resistance, of saying "no" to power, and as with the parrhesiastes, accepting the consequences. Dick's constant validation of the everyday person, Joe Chip, Mr Takomi, or Ragle Gumm, or of the destruction of the environment serve to construct connections.

PART III

Case Studies in the Production of Cyberspace

CHAPTER 6

Disciplinary Cyberspaces: Security and Surveillance

The earth spins on its axis
One man struggle while another relaxes
As a child's silent prayer my hope hides in disguise
While satellites and cameras watch from the skies
Massive Attack, Hymn of the Big Wheel

In this chapter I focus on how the subject is formed or "made up" (Hacking, 2002) within disciplinary spaces. I do this by focusing on the digital cartographic representation of criminality; of how subjects are processed in crime maps; and then how that information leads to a whole set of surveillant practices and outcomes in physical spaces, as well as the need for techniques of data collection and distribution. I will try to expand the concept of "cyberspace" by analyzing it not just in terms of the Internet, but in terms of digitality and the virtual as outlined in Chapter 1.

Security has now become the dominant framework for understanding the modern world in the United States. It is enshrined as both law (e.g., the USA Patriot Act 2001) and official doctrine and policy (i.e., the National Security Strategy of the USA 2002). Significant resources are being dedicated to maintaining and increasing security, as well as the deployment of technologies to identify and pre-empt possible threats to security. Many of these developments represent important shifts in policy and could mark a new political era which abandons the policy of containment and deterrence of the Cold War for one of threat assessment and unilateral action where necessary. Although security has never been unimportant to nations, what is new is the almost complete subsumption of politics under this heading.

Digital mapping and GIS play an important function in the practice of security. On the one hand they are technologies which are put into practice on an everyday basis, and on the other they are the means by which particular ways of thinking (rationalities) are enabled. I am especially interested in how a rationality of "carto-security" is being established, that is the deployment of mapping and GIS for the ends of security. What are the effects of such a rationality? What kinds of possibilities does it enable or close off for human beings? How does it produce consequences that are embedded in everyday life? Are these consequences being taken into account? If not, are there other ways in which mapping and GIS can be productive and useful? These are not easy questions to answer. But I will try to do so by taking a case study that offers insight, that is crime-mapping.

Crime analysis is one of the largest applications of GIS and digital mapping within the framework of security. Who uses GIS to analyze crime? According to the Department of Justice's Mapping and Analysis for Public Safety (MAPS, formerly the Crime Mapping Research Center or CMRC) a broad range of entities use crime-mapping. Primary users are police departments who have swiftly been taking up GIS as a crime-fighting tool. Other users include local, state, and federal government, and local communities (especially those with experience of public participation GIS). But the broader security field may be almost unlimited. Many companies now offer equipment that promises to track or tag trucks (fleet management), people (parolees, children, "impaired seniors"), or pets. All these activities are deeply geographic and usually require sophisticated mapping solutions. Following the attacks on the World Trade Center and the Pentagon, some security experts have been estimating the size of the "security market" as worth anything up to $100 billion by 2010.

Are we safe? Much of today's discussion about security in the United States from a geographic standpoint is clearly meant to improve our safety. From this perspective it's hard to be "against" security. The case for mapping and GIS was made apparent very soon after September 11 when the *New York Times* and NBC published several powerful images by the Center for the Analysis and Research of Spatial Information (CARSI) at Hunter College. These LIDAR (light detection and ranging) maps showed a before and after view of the World Trade Center in 3D. In that first week, LIDAR images were one of the few ways to penetrate the smoke and dust, and were collected on a daily basis by airplane (satellite imagery was also collected by Space Imaging using the IKONOS 1-meter resolution satellite system). The CARSI lab had previously produced a 3D map of the city, called the "NYCmap" based on building heights integrated into high-resolution digital orthophotoquads (DOQs). When placed side to side, the jagged hole around "Ground Zero" became eloquently visible. In the following months, ESRI offered a series of seminars around the country on how GIS could assist in emergency prevention and response, published white papers, produced a CD-ROM of material on security and established a website.[1] The professional geography community meanwhile launched an NSF-funded workshop on "geographical dimensions to terrorism" and established a list of priority action and research items. The first priority action item listed is to "Establish a distributed national geospatial infrastructure as a foundation for homeland security" (Cutter, Richardson, and Wilbanks, 2002, p. 2). Geography, and especially GIS, remote sensing, and digital mapping, were integral parts of the security effort in order to compile, integrate, and distribute spatially referenced information. But September 11 was a continuation of a concern with security and disciplinary space, a tremendous heightening, but not something new. Many people thought following the attacks that September 11 would "change everything". It would seem that a better way to understand it is as a deepening and intensification of trends already in place, at least as far as disciplinary spaces go.

From a digital mapping standpoint I shall make a fairly straightforward argument: that by constituting people and space as at-risk resources to be managed it is all too easy to slide into a discourse of risk and normalization (what is a normal and what is an abnormal risk), and statistics (actuary tables, censuses, polls, and thematic maps). Such a discourse is an extremely limited and impoverished understanding of lived human experience, especially as it resurrects yet again the effects of normalization. If this is coupled, as it often is, with the conception of threats from anywhere and to anywhere, and hence the need for an equally extensive surveillance system, then we should be able to recognize the need to think critically about the role of geography as a discipline and mapping as a practice in today's efforts to improve security. In Foucault's phrase the effects of security reach "omnes et singulatim", or all and each one individually (Foucault, 1988b).

Criminality is a threat to security, and has been understood as such for a long time. But has it always been understood in the same way? In a lecture to the Law and Psychiatry Symposium in Toronto in 1978 – as Foucault was shifting his focus from technologies of domination to technologies of the self – he argued that the judicial system underwent an important shift in its approach to criminality. Prior to the legal reforms of the eighteenth and early nineteenth centuries the law focused on the nature of the crime committed, the evidence of guilt or innocence, and the system of penalties which would be applied. In other words: crime and punishment. The person of the criminal was important only insofar as he or she was the individual to which the crime would be attributed. With the reforms, this hierarchy was reversed, the crime was merely an indicator of something more significant: the "dangerous individual" (Foucault, 1988a). The law was now interested in the potential danger of the individual: "the idea of *dangerousness* meant that the individual must be considered by society at the level of his potentialities, and not at the level of his actions; not at the level of the actual violations of an actual law, but *at the level of the behavioral potentialities they represented*" (Foucault, 2000a, p. 57, original emphasis). It became necessary to know the criminal as a subject, to have him either confess and provide an account of himself, or to provide those aspects for him if he was unwilling to do so himself. Thus punitive responses had to be appropriately tailored to the perceived threat of the individual, to who the criminal actually was (which at that time highlighted medical or psychological knowledge). This explains the emergence of the criminal psychiatrist as an expert figure who could interpret the behavior and actions of the criminal.

The criminal is obliged to confess his or her territorial dangerousness, either directly, or through a panoply of techniques for computer crime-mapping. The contemporary concern with crime has given rise to a whole mode of digital spatial analysis, such as start-up companies who analyze spatial criminality (called geographic profiling or geo-profiling), through statistical mapping methods, spatial surveillance devices which allow more and more criminals to be paroled into the community as long as they stay within certain spatial territories or the journey to work corridor, and specialized computer software (such as "CrimeStat") designed to aid in this analysis. Most urban police departments now have a GIS analyst equipped with ArcView or MapInfo, read the Crime Mapping Listserv, watch GIS analysis of criminality on the TV show *The District* (linked to from ESRI's website), and so on. Somewhat surprisingly given these important facts, there has to date been little sustained analysis of the effects of cyberspatial surveillance of the criminal through digital cartography and GIS (though see Graham, 1999 on CCTVs, and Monmonier, 2002a). What I think has been especially overlooked is the question posed by Foucault of the nature of how subjects are produced or made up within disciplinary spaces – in other words, the whole production of disciplinary cyberspaces and how they act to subjectify the individual.

EARLY APPLICATIONS OF CRIME-MAPPING

As I mentioned above, a significant development in the judicial system was the emergence during the nineteenth century of the criminal as someone who had to be *understood* in order to be worked on and normalized. In addition to both the crime and its punishment it became necessary to know why the criminal acted as he or she did. This understanding would produce the context in which the individual's dangerousness could be deduced. Dangerousness is a measure of the degree of risk or threat posed. What Foucault suggests therefore is that by pursuing a discourse of understanding the criminal, the issue is framed as one of *risk* and *threat*. One may see this as being a risk or being at-risk. How can risk be assessed and compared from one place to another? Are some things more at-risk than others? In order to answer these questions some people at the beginning of the nineteenth century turned to the emerging practices of statistics and statistical thematic mapping. It is not, of course, a coincidence that statistics and risk emerged hand in hand. As Hacking (1990) has shown, the surfeit of numbers, the invention of the normal distribution (by Adolphe Quételet and hence of normal man), and the application of statistics to social problems (as "moral statistics") and policy-making were all bound up with what Hacking calls "the taming of chance", that is the application of statistical regularities to social life. Hacking's work has been inspired by Foucault's bio-politics and the governance of populations. Hannah (2000) similarly applied it to the governance of territory in the late nineteenth century. But no one has yet tackled the issue of subjectivity and space through the cartographic production of knowledge.

The first efforts at mapping crime were a preventative measure – in order to get a better description of where crime was occurring and what kinds of crime were being committed. From this it became possible to formally differentiate different neighborhoods of the city and to classify space in terms of safety: crime-ridden; marginal (either in decline or improving); crime-free. The crime map became an important means of constructing the city. The application of crime-mapping within police departments is quite recent (the 1980s). Weisburd and McEwen (1997) argue that few departments had the ability or patience of think about crime spatially, and it was only with the availability of computers that could record locations, respond to emergency calls accurately, and process and track offenders that the spatial distribution of crime became manageable. But the first (hand-drawn) maps of crime were actually produced 130 years earlier, in 1829.

Thematic (statistical) mapping as a whole, and the choropleth map in particular, appears relatively late in the historical record, but precisely at a critical juncture in the development of the state. Few thematic maps on social subjects appeared before the early nineteenth century, whereas maps of the physical world (ocean currents, geology, magnetic phenomena) date from about 1650 or even earlier (Robinson, 1971, dates the first "iso" map to 1584, that is a map showing lines of equal value). Later, in his definitive history of early thematic mapping, Robinson (1982) suggests several reasons for the cartographic retardation of the social sciences compared with the physical sciences, of which the most interesting from our perspective is the sheer lack of numerical data for the *population as a whole*, rather than partial (local or regional) counts. The turnaround is coincident with the beginning of the nineteenth century, when for example the first French and English censuses were taken (1801).[2] To this was added, after the 1820s, the new analytics of probability theory by the Belgian statistician Adolphe Quételet, who invented the normal distribution curve. Quételet was concerned with social upheavals during the 1830s and centered his analysis of social variation around *l'homme moyen*, or the average man. Indeed, this reduction of variation to a *norm* is central to the practice of thematic mapping.

The statisticians Adriano Balbi and André-Michel Guerry, who produced the first maps of crime in 1829, used the then newly-invented technique of choropleth mapping. Today the choropleth is an extremely popular type of map; it is often the default option on mapping and GIS programs, it is used extensively in newspapers and other popular media, and to my certain knowledge is all too often chosen by students in cartography classes. But it too was invented, by a Frenchman, Baron Charles Dupin, in 1826. Dupin, Balbi, and Guerry were all interested in how to understand France, its territory and resources (Dupin's map was of educational levels and sought to classify France into "enlightened" and "unenlightened" regions). When Balbi and Guerry made their crime maps, they related crime to education using crime data from the new *Compte Général de l'Administration de la Justice Criminelle* for the years 1825–27 and the French census (Robinson, 1982; Porter, 1986). This data report was an early forerunner of crime statistics such as the annual Uniform Crime Report (UCR) published by the FBI. The role played by the national census cannot be understated. It certainly aided the development of statistics as a science, as did the larger activities of government data collection which took off in Europe and America in the early 1800s (Edwards, 1969).

There had been many previous attempts to portray statistical data in graphic form, but these were sporadic. According to Funkhouser (1936, 1938) the earliest may be a tenth- or eleventh-century line graph of planetary orbits (also reproduced in Tufte, 1983, p. 28). This form of plotting data disappeared for another 800 years. By the seventeenth century there were attempts to understand economic data in tabular form, for example by Charles Davenant, Johann Peter Süssmilch, and Sir William Petty (described by Funkhouser as "the Quételet of his day", 1938, p. 279 and who invented the term "political arithmetic"). Joseph Priestley published a bar chart of biographical and historical events using an early version of the timeline (1765). But the man who surpassed them all, while (sometimes) acknowledging and adapting these earlier attempts, was William Playfair, who developed the pie chart, bar graph, proportional symbol, and modern line graph. Playfair published his (somewhat misnamed) *Commercial and Political Atlas* in 1786 (two later editions in 1887 and 1801 contained no maps) and his *Statistical Breviary*, also in 1801, and another significant work towards the end of his life, the *Letter on our Agricultural Distresses* (1821). A small digression on Playfair is worthwhile.

It is an interesting question why Playfair, who was neither a scientist nor a scholar but rather a political pamphleteer, should be the first to invent these statistical graphs (Biderman, 1990; Spence, 2000). And why might it be the case that he developed them when he did (late eighteenth and early nineteenth century)? After all, Descartes had already provided the grounds for geometric and mathematic graphing 150 years earlier. There had been scattered usages here and there (Priestley, Graunt). But Playfair was the first to consistently use statistical graphics. We might also ask why Playfair chose economic subjects to graph. In fact there is one answer to both these questions (why then? why economic data?). Statistical graphs of the economy were not needed prior to the early nineteenth century. At the time of Playfair's intellectual efforts, however, both statistics and the application of them to social problems (as moral statistics) became possible. A new understanding, or rationality, of how to run the country was in operation. The model based on sovereignty was giving way to one based on management of people, territory, and resources. As for Playfair himself, he was a member of the Edinburgh Enlightenment, he published an edited edition of Adam Smith's *The Wealth of Nations* (1805), and when he was a boy his brother John had him record temperatures over time by drawing lines on a divided scale (Biderman, 1990, p. 9). When Playfair was seventeen he began working in Birmingham with the inventor James Watt as his personal assistant and draftsman. This apprenticeship is significant. In Watt's day it was very difficult to record the continuous state of an instrument in operation (no databases or readouts), and so Watt used a physical recording device which scribed a line on a piece of paper or smoked glass (the Watt indicator). Thus by physical necessity graphs were produced (Hankins, 1999). Playfair also met Joseph Priestley at this time. So Playfair had every opportunity of circumstance to develop and apply his techniques (and apparently of character too – among his life experiences are blackmail, the storming of the Bastille, a failed bank business, and self-aggrandization). However, a purely personal or psychological explanation for Playfair's innovations (Biderman, 1990; Spence and Wainer, 1997) is not sufficient. If not Playfair, then Humboldt, Mayhew, Lambert, Minard . . . or Balbi and Guerry.

Balbi and Guerry expected that the areas of France with lower educational levels would correlate with a higher incidence of crime. In this, they were rudely shocked: it was just the opposite. Where education was high, so was crime. As Hacking portrays it:

> Such a conclusion was sensational. Paris saw itself as being in the grip of a terrible crime wave. Ask a New Yorker of today [1990] about muggings, then double the fear: that was how Parisians felt. The [illustrated] police gazettes, rich in reports of crimes, were taken in weekly, and were the sources of best selling novels like those of Eugène Sue. Naturally one supposed that the degeneracy and ignorance of the working classes was the source of their criminal propensity . . .
>
> (Hacking 1990, p. 78)

Hacking's description is evocative and insightful in highlighting how the first statistical descriptions created challenges for the political economists, carto-statisticians, and lawyers (Dupin, Balbi, and Guerry respectively) who attempted to get a handle on society. After the initial results came in, Balbi for one decided that it was essential to consider a whole raft of other factors that might influence the results. Many of these were geographical, such as nearness to the coast, population density, the influence of borders on the data, and so on. Crime had become a problematic of statistical space.

Guerry and Quételet were also astounded that crime could vary so little from year to year. In 1833 Guerry remarked that "if we consider now the infinite number of circumstances that can cause the commission of crime, . . . we will not know how to conceive that in the end result, their conjunction leads to such constant effects" (quoted in Porter, 1986, p. 49). Was this regularity a statistical artifact, or some indication of nature's harmoniousness, God's wisdom (and what then of free will?). Debates raged in a way we would find very peculiar today. Statistics is the study of populations and their regularities, but suddenly individuals (and individuality) was at stake. If populations and social phenomena obeyed regularities as was revealed by Guerry, Quételet and Laplace (who showed the number of dead letters in the Paris post office remained constant from year to year), what then for individual human will? Quételet started to speak of the average man, *l'homme moyen*, and it became harder to be so confident that statistical regularities in crime and theft were part of God's plan. Different statisticians approached the problem differently, as Hacking (1990) has detailed. Some, like Quételet, denied that it was necessary to abandon reference to a divine plan. But what could not be denied is that society, population, territory, government was now a matter, indeed even a problematic, of social statistics.

Yet Hacking, like many historians of social statistics, does not spend much time discussing their cartographic history, once again overlooking the critical importance of the production of spatial knowledge through mapping. On the other hand, where cartographers, historians, and scholars of crime-mapping have traced the development of thematic maps (e.g., Robinson, 1982; Konvitz, 1987; Dent, 2000) they do so without Hacking's (or Foucault's) discussion of governmental rationalities. Is it possible to productively conjoin these two literatures in a way that will offer insight into the production of subjectivity in disciplinary spaces?

GOVERNMENTALITY

Statistics, and their graphical depiction in maps, were essential aspects in the state's control and regulation of its population as it pursued a governmental rationality or governmentality. Thematic maps require national mapping programs in order to determine the internal boundaries and borders of the country, and national censuses to collect the statistical data. Foucault's notion of governmentality highlights the historical emergence in Western modernity of a new "governmental" mode of politics and power. Sovereignty, with its absolute right to decide life and death, and the divine laws of Christianity are replaced by a concern to foster life or to disallow it to the point of death in capital punishment (Foucault, 1978, p. 138).[3] Foucault traced two specific aspects of this shift: one centered on the body through discipline and optimization of its capabilities – an *anatomo-politics* – and one centered on the population as a "species body" or a *bio-politics* (Foucault, 1978, p. 139; see also pp. 25–26; Foucault, 2000b, pp. 215–18). Where discipline emerged during the seventeenth century, bio-politics followed during the eighteenth.

Foucault claimed that there was an explosion of interest in the problem of government from the late sixteenth century on (focusing on government of the self and pastoral/religious care), to texts about the state and the government of others. The most emblematic is the shift between Machiavelli's *The Prince* (1513) and the *Miroir Politique* by Guillaume de la Perrière (1555, English edition 1598). Machiavelli's book is about how the prince can hang on to his territory and resist threats, but who is always "external" as it were (separate from the state itself). In Perrière Foucault detected nascent governmental concerns, which have a "multiplicity and immanence" (2000b, p. 206) entirely lacking in Machiavelli's worldview. They cover everything, from family to economy. By the time of Rousseau's article on political economy (1755, which went into Diderot's *Encyclopédie*) the art of government "means exercising toward its inhabitants, and the wealth and behavior of each and all, a form of surveillance and control as attentive as that of the head of a family" (Foucault, 2000b, p. 207).

This sentence neatly summarizes the issues at stake. First, there is a concern to manage (economize) such things as wealth, health, and certain behaviors that affect the population (i.e., education, birth rates, literacy, and so on). Management, husbandry, and regulation (e.g., with the "police") constituted people and territory as a precious resource to be protected from threat and depletion. Second, these concerns could be directed at two levels, the level of the body as the anatomo-politics or the level of the population as the bio-politics ("each and all" or more classically "omnes et singulatim" – Foucault, 1988b). Finally, techniques of observation were required in order to practice the management. If the state did not know what was going on, it would be too hard to govern effectively, pinpoint problems or threats, and perform the required resource management. From all this then a distinct set of knowledges and of practices (police, surveillance, data collection) were to arise. This governmental rationality is the context in which I interpret digital crime-mapping today.

DIGITAL CRIME-MAPPING AND SURVEILLANCE

Like the original maps of Balbi and Guerry, the modern crime map is tied to the rise of social statistics such as the FBI Uniform Crime Report (UCR) which has been collected since the 1930s due to the rise of organized crime at that time. The national censuses in the USA and Europe also provided a resource which could be graphed, and less frequently (at first) mapped. Today we see a hyper-extension of these developments in the surveillant systems deployed by the police to monitor and check residents as they go about their daily business. These include closed-circuit TV (CCTV), digital face-matching such as the one in Tampa Florida (Canedy, 2001), and the FBI's new DNA database, the Combined DNA Index System (CODIS), which since 2000 has been authorized to collect DNA data from persons convicted of violent crimes (FBI, 2001, p. 1). In 2001, the KlaasKids Foundation advocated a similar act for the state of California for burglary, robbery, carjacking, and arson The discourse surrounding this call was situated as one of "public safety" (KlaasKids, 2001).[4]

Graham (1998) discusses some implications of regulating space by what he calls "surveillant simulation" (i.e., digital surveillance or "control at a distance" such as electronic tagging – see Bloomfield, 2001).

Graham highlights four cases of surveillance: as social control especially of criminality; in and around consumption; differential deployment over space (transport informatics); and the utility industry. To this list we could add others, such as surveillance in the workplace (a practice established at least as far back as the nineteenth century) or in the pursuit of leisure activities (e.g., the Visionics face recognition software and CCTVs at stadium sporting events) or walking down the street (one estimate is that there are over 2.5 million surveillance cameras in Britain – see Rosen, 2001). Graham argues that CCTV is used to target individuals who do not "belong" or are abnormal in their behaviors and thus are more dangerous and likely to commit criminal offenses.

Graham's discussion is very useful in identifying some of the main concerns with surveillance technology. However, we might register a point of caution concerning the idea of social control. As far as Foucault's work is concerned, "social control" should not be interpreted as a condition of total domination over life. Foucault emphasized *discipline* and the governmental *management* of a problem.

Although Foucault does not use the word *geo*-surveillance, the necessity for it is built right into his descriptions of modern society. Bodies may also be disciplined by the clever design of spaces that would make surveillance easier (the panopticon can be cited here, but there are a myriad of examples from the half-doors of toilets so that teachers can see the legs and feet of students, to the spatial partitioning of the plague town, to camgirl and webcam competitions in which participants reveal their bodies and living spaces).

There are many techniques of geo-surveillance employed in crime-mapping. One set of techniques revolves around those who have already offended and are in the criminal system. At year-end 2001 there were about 6.5 million people classified as offenders in the USA. Those that are incarcerated numbered 1.96 million. However, there are also about 4.66 million people on probation or parole (3.93 m and 0.73 m respectively) which usually involves some degree of self-reporting (Bureau of Justice Statistics, 2002). A small number of people not in local or state jails are supervised by other techniques, including community service, work release, weekend reporting, electronic monitoring, and other alternative programs. Jail populations have steadily increased during the 1990s (by about 30 percent between 1995 and 2001), prompting increased efforts to supervise this population outside the facility (Aungles and Cook, 1994).[5] During the same period the percentage of people supervised outside jail rose from 6.4 to 10.0 percent (34,869 to 70,804). However, the number of people at the local level supervised by electronic monitoring has remained the same for several years, at about 10,000. This number is less than 15 percent of those monitored outside jails, and a tiny 1.4 percent of the total jail population.[6]

In order to administer the criminal's dangerousness a new field of *expertise* is required. Similarly, in crime-mapping, we also encounter a whole regime of expertise through which spatial dangerousness is assessed. Although in some cases these experts are part of the judicial system itself (typically GIS researchers or police staff in police departments across the United States) there are also a set of extra-judicial ones. The latter include companies that monitor offenders on parole or probation via electronic monitoring, purveyors of criminal statistical GIS software such as "Crimestat", organizers of crime-mapping conferences, academics, providers of monitoring anklets and security systems for surveillance, website authors producing parolee mapping, neighborhood associations agitating for offenders to be registered, and so on. These experts are part of a new set of power-knowledge relations for assessing spatial dangerousness.[7]

From this domain of expertise also emerges the need for a new set of *techniques*. What are the techniques for spatially subjectifying the criminal in cyberspace? They are GIS and digital mapping, often but not necessarily deployed through the Internet. The specific techniques are known in the field as "geo-profiling", coined after the idea of psychological profiling. The theory of geo-profiling was developed by Kim Rossmo in 1995, and has since been implemented in software which can make a predictive surface of a criminal's location.[8] In Rossmo's words geo-profiling is "an investigative methodology that uses the locations of a connected series of crime to determine the most probably area of offender residence" (Rossmo, 2000, p. 1). Geo-profiling is based on the concept of offender behavioral profiling (made famous in the movie *Silence of the Lambs*) that produces a prediction of the offender's characteristics, in this case his or her home location. Rossmo uses well-known principles of geography, such as Tobler's first law and the BOF (birds of a feather) principle that crime is committed by people near (but not too near) their own homes. Rossmo claims that with 5–6 incidents traceable to one person, his software can reduce the search area by up to 90 percent.

In geographic profiling the expert is able to identify spatially dangerous locations (the target of the criminal, taken individually or altogether). For example, a series of crimes with a certain characteristic implying they were performed by the same person could be mapped, and with special statistical techniques (such as spatial autocorrelation) either the next target area can be predicted, or the location of the criminal's residence narrowed down. This knowledge is produced from the criminal's spatial

Sexual offenders and the production of disciplinary online spaces

The classic case of the disciplinary society operating in cyberspace is that of sex offender registration and mapping. In this case, they are people who have admitted their guilt or been found guilty by the judicial system, and who have served a sentence in prison. In theory, the judicial system is concerned with determining guilt and weighing punishment which is then served. When offenders are required to register, either by the state or the government, or are tracked and mapped privately by community organizations after their sentence, the concern is *not* with catching the person who commits a crime, but with *assessing who might be dangerous in the future*.

During 2000, the UK saw a number of public protests stirred up by a media campaign of "naming and shaming". *The Times* noted six and even four-year-olds outfitted in T-shirts with phrases such as "Kill the paedophiles" and "paedophiles are scum" scrawled in red lipstick on them. A six-year-old girl was "surrounded by a mob of baying, screaming mothers, many pushing prams, young men drinking beer and about 80 other children, all chanting: 'Sex case, sex case, sex case, hang him, hang him, hang him.' Foucault observes the irony in constantly speaking about children's sexuality (in the eighteenth century) where the ostensible purpose was to "prevent children from having a sexuality" (Foucault, 1980d, p. 120) because these discourses had an almost opposite effect, which was

> To din it into parents' heads that their children's sex constituted a fundamental problem in terms of their educational responsibilities, and to din it into children's heads that their relationships with their own body and their own sex was to be a fundamental problem as far as *they* were concerned; and this had the consequence of sexually exciting the bodies of children while at the same time fixing the parental gaze and vigilance on the peril of infantile sexuality . . . "sexuality" is far more of a positive product of power than power was ever repression of sexuality. (1980d, p. 120)

behavior and "journey to crime" models which can be derived from the archives of criminal behavior over space (see box).

Several technologies have been developed to monitor such offenders. A common technology is an ankle bracelet or tag which emits an RF radio signal that can be detected by a device in the home linked to the phone system. The receiver can call the monitoring center if the anklet's signal is no longer detected (i.e., if the offender has left the house or gone out of range). The range can be adjusted to suit the needs of the case. According to one US company, Behavioral Intervention (BI, Inc.), "today, approximately seven out of ten offenders are not behind bars, but reside within our neighborhoods" (BI, 2002, n.p.), a figure they use to promote their HomeGuard™ monitoring system. By implying that offenders have already interpenetrated "our" own living spaces, the need for a commensurately penetrative surveillance system is made paramount. This mentality is exactly that of the Cold War and the Red scares of the 1950s. Johnson, for example, has suggested that when the Russian NKVD cables were broken on one occasion (the Venona transcripts) in 1948 they "alerted the West to the doings of Guy Burgess and all manner of others . . . the shock was huge: the enemy was within the gates – and in great numbers. Venona almost certainly led to McCarthyism" (Johnson, 2002, p. 16). The mentality of threat and dangerousness is *itself* not without risk, in the 1950s engendering a society of control and suspicion. We can call this the risks of security.

A more advanced approach is to use GPS. It too is often based on an anklet worn by the offender which can receive GPS signals and transmit its location (through the cell phone system) to the company's monitoring center. In Iowa, for example, the police have required some offenders to wear a device from a company called iSecureTrac which tracks individuals with GPS and transmits the location to the Web. Furthermore, this monitoring is very geographically specific and can incorporate exclusion zones:

> Each map is tailored for a specific parolee. A map can show, for instance, areas where a paroled pedophile must remain clear of – such as a school – when going to and from an off-site counseling session. When the parolee returns to the halfway house, he places the device in a docking station, which transmits the data to iSecureTrac.
>
> (Chabrow, 2002)

In a recent interview Mark Monmonier noted that there are even more sophisticated proposals, such as that of Digital Angel (Monmonier, 2002b). Digital Angel is a Minnesota-based company that provides various monitoring devices that can be tracked by GPS. Monmonier mentions one such device that can be inserted under the skin of children or pets, and can deliver an electric shock or "e-pinch" if the person transgresses certain predetermined borders (e.g., a parole zone). Digital Angel has been quoted as estimating the potential size of the "personal safeguard technology" market at $70 billion (Wireless NewsFactor, 2001). Even if

Gun - Drug Connection
City of Atlanta 2001

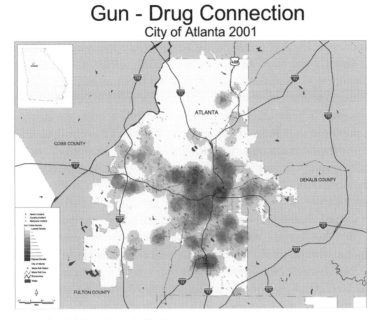

Figure 6.1   Spatial detection of offenders.

Source: Chris Gundry/Georgia Data Center (GADC). Used with permission.

this is an exaggeration, the potential applications are diverse, including commercial fleet management, agriculture (cattle), medical ("cognitively impaired seniors"), children and pets, as well as the still small but growing field of offender monitoring. In concept, this is not new, and one can trace similar proposals to at least the early 1970s (see, for example, Ingraham and Smith, 1972, who proposed the use of electronic devices to location-ally monitor parolees). More generally, proposals for offender marking, for example with indelible ink, date back to the nineteenth century.

In another example, the Omega Corp. has provided an online crime-mapping solution called CrimeView. This technology is based on ESRI's Internet Map Server (ArcIMS) and allows the usual range of queries which are possible in GIS, including buffering (e.g., locate all sex offend-ers within half a mile of a school), query building, various layer overlays (e.g., parolee and school locations), and "hotspotting" for offences (e.g., traffic violations). These crime maps enable geo–profiling to isolate behavior which does not conform to the norm, or to prevent norms from being transgressed by people deemed to be dangerous. However, hotspot-ting maps are now giving way to more sophisticated spatial analyses such as those pioneered by Rossmo mentioned above that permit spatial buffering and spatial detection of offenders (see Figure 6.1 and box).

A crime-mapping discussion

In order to learn more about issues of crime-mapping I subscribed for about two years to the public forum crime-mapping listserv ("Crimemap", sponsored by the National Institute of Justice). Among the usual announcements of conferences and job openings for the "experts" (I reproduced one from November 2000 earlier in this chapter) the vast majority of discussion centered around techniques for improving the accuracy ("hit rate") of crime-mapping; whether to use choropleth or isopleth maps, how to georeference the often vague locational data recorded by police officers, and the ability of crime maps to truly show the situation on the ground (or whether they were too vague). Perhaps the most interesting discussion centered around making these crime maps publicly available, where presumably not only the general public but also potential criminals could access them and modify their criminal behaviour (May, 2001). The general impression I received was one of a group of hard-working and dedicated researchers trying to grapple with technology (the most frequent discussions centered around problems with ArcView) and with a deep concern to make their techniques work more satisfactorily. Certain people tended to provide answers over and over again, and it was clear that even among this relatively small professional group there were leaders with more experience. Almost none of them seem to be involved in electronic tagging, rather they were modeling, predicting, or trying to solve criminal behavior in the manner outlined by Rossmo (who briefly joined in the discussion in April 2001).

What I didn't see was any kind of insight into what they were doing and the effects it has on characterizing people as potentially dangerous, as at–risk resources under threat. Nor was there any questioning of surveillance (or geo-profiling) or where the data came from that they used to model behavior.

Since in Florida law it is illegal for a sex offender to live within 1,000 feet of a school the "Enforcer" software, which consists of digital mapping tools, can locate and flag these individuals. Using transparent overlays of map information from Florida and other states, the system can integrate data from many sources and keep it relatively up to date. Compared to other counties in Florida, Pinella County no longer has to measure distances with wheel measurers (like the ones often used at accident sites). The other interesting aspect of this computer mapping and tracking system is that is is justified on the grounds of the benefits it offers. In a news story, Lt Tom Evans of Pinella County was asked about

How long have you been serving your time?

In my research for this chapter I originally had links to a well-known crime and police information site called APBNews.com. This site gathered news of all sorts relating to crime, the police, missing people, and jurisprudence. APBNews.com had financial problems in 2000, went into bankruptcy, and was eventually bought out by a dot.com startup called SafetyTips, which itself went bankrupt. (Full disclosure: I once did some work for SafetyTips and was owed a substantial sum at the time of their bankruptcy.)

Since APBNews is now off the air, I decided to replace my links by accessing the Internet Archive, which has been saving websites since 1996. Although I shouldn't have been, I was surprised to find that the archive still stored the names, addresses, and in many cases the photographs of offenders that it had obtained from visits to police department websites. The Atlanta Police Department (APD), for example, has a "naming and shaming" page they dramatically called "Busted!!" with a little picture of a grim felon staring out from behind prison bars. This page names and shames offenders in only two categories: sex crimes and drug crimes. I was able to go back to October 25, 1996 and get names of these offenders, even though they are only charged with a crime and may or may not actually be guilty.

This is not to criticize the Internet Archive, without which Internet research would be much harder. However, it *is* an unintended consequence of putting information out publicly: once it's out, it's out; and it's as if you will continue to serve your time forever.

the possible "Big Brother" implications of integrating all this information. He replied that "Hopefully . . . we're going to offer them a safer community" (APBNews, 2000 – see box).

In some communities the practice of profiling has been challenged. After a series of high-profile incidents on the New Jersey turnpike in which African-American drivers were disproportionately stopped by the highway patrol, it was charged that the police were stopping blacks because of who they were, not because of their actual behavior.[9] In other words the judgement was made on the basis of potential dangerousness rather than actual offenses being committed (i.e., searches were made without probable cause). This racial profiling was very controversial in the black community and led some police forces to issue new procedures. Yet it *is* the case that profiling depends for its success on specifically targeting certain groups or behaviors because they are deemed to be (or empirically actually are) more dangerous (for example, sex offenders walking near schools) even if no crime has taken place. Whether or not blacks are statistically more likely to be transporting drugs – and this is disputed by Rutgers Law School professor Sherry Colb (2001) – crime is being understood as deviancy from a norm. It is this normative rationality that I wish to underline here, and how statistics are used to assess it, and not the accuracy of the statistics per se. To date terrorist (racial) profiling has not been problematized in the same way as has racial profiling, presumably because we grant that the government has a compelling urgent interest in stopping terrorism. But the principle behind it is the same.

The need for blanket monitoring and surveillance arises because of the perceived ubiquity of the threat. Often we don't know in advance where the risk or the danger will be. As the Fire Chief of the Livermore/ Pleasanton Fire Department put it on the "Homeland Security" CD-ROM distributed by ESRI:

> I think now that everyone's reminded that anytime, anywhere, a significant catastrophic event can occur. An industrial accident, internal sabotage, external terrorism, a bad weather that hasn't come in a hundred years, and that our citizens expect everybody to be prepared for that.
>
> (ESRI, 2002)

In other words when ordinary (non-military) places such as the World Trade Center are attacked it confirms that we are already "inside" the threat zone and therefore must constantly map everything. To put it another way, the enemy is within the gates. Thus the new reality of the threat: it is everywhere and so must be the surveillance.[10]

The same reasoning applies to the Bush administration's controversial plans for TIPS (the Terrorism Information and Prevention System) that was proposed in early 2002. In this plan, citizens and workers who often go into residential neighborhoods (postal workers, cable TV installers, truck drivers) would be recruited to call a government hotline if they saw suspicious activity. The idea was to benefit from as many as a million sources of surveillance in ten pilot cities (these cities were never specified). An $800,000 call center was to be funded by the Department of Justice at the National White Collar Crime Center (NW3C), a Congressionally-funded NGO that coordinates information about Internet and high-tech crime founded in 1980. The plan ran into immediate trouble and the House Select Committee on Homeland Security promised to insert language into the Homeland Security Department bill to block it. Although the Justice Department continued to support TIPS as an initiative of the

"Citizen Corps" it now no longer plans to issue the hotline number to any worker "in contact" with homes or private property. As late as August 2002 the government TIPS website asserted it would be implemented in Fall 2002, arguing that:

> Industry groups have looked to the Justice Department to offer a reliable and cost effective system that their workers could use to report information to state, local, and federal law enforcement agencies about unusual activities they might observe in the normal course of their daily routines.
>
> (Citizen Corps, 2002)

Although TIPS was finally dropped, other terrorist information hotlines, such as the FBI's Terrorist Tipline, remain in operation (as well as Amber Alerts about child abductions, Coast Watch, Highway Watch, and River Watch, for reporting chemical or biological spills, the ATF Hotline for reporting suspicious firearm activity, and even the Treasury Department's FinCEN for reporting financial crimes). In November 2002, the *Washington Post* and *New York Times* revealed another surveillance initiative at the Pentagon called Total Information Awareness (TIA). TIA would use computer data mining techniques on everybody's daily transactions such as credit card purchases, telephone calls, and Internet traffic like email in order to discover at-risk behaviors. Finally, the "Uniting and Strengthening America by Providing Appropriate Tools Required to Intercept and Obstruct Terrorism" Act of 2001 (the USA Patriot Act or USAPA) also widely broadened existing statutes permitting electronic surveillance. According to an analysis by the Electronic Freedom Foundation (EFF) USAPA "expands all four traditional tools of surveillance – wiretaps, search warrants, pen/trap orders and subpoenas" (Electronic Freedom Foundation, 2001) and made it legal to search homes without a warrant or install surveillance devices in people's homes without notification until long afterwards (see HR 3162, §§202, 210, 213).[11] We have thus reached an analogous situation to that faced by the citizens of Paris in 1829 when faced with the Balbi and Guerry crime maps: we fear crime and threats to our security from everywhere and must deploy surveillance as much as possible.

IS PRIVACY THE ISSUE?

We often characterize the issue of surveillance – as exemplified by crime-mapping and geo-surveillance – as an infringement of our privacy. Or we examine the effect increasing surveillance has on (what's left of) privacy. Monmonier, for example, in his pioneering book on geo-surveillance largely sees surveillance in these terms (Monmonier, 2002). His subtitle is "surveillance technologies and the future of privacy". The implication is that surveillance and privacy are the terms in opposition to each other. The theme percolates many of the classic texts on surveillance. Burnham (1984), for example, states that "the gradual erosion of privacy is not just the unimportant imaginings of fastidious liberals. Rather, the loss of privacy is a key symptom of one of the fundamental social problems of our age: the growing power of large public and private institutions in relation to the individual citizen" (p. 9). The "power" referred to consists of corporate-legal data collection techniques and practices of observation.

As the computer age progressed, more attention was paid to digital or electronic surveillance. Lyon (1994) gives a quick history of the emergence of privacy as a desirable state and human right (declared by the UN in 1948). Alderman and Kennedy (1997), two attorneys, give details about some of the types of privacy cases that have helped establish case law (beginning with the fact that the word privacy does not actually appear in the US Constitution). They feel that awareness of privacy is on the rise in the USA: "people are realizing that they are vulnerable to invasions of privacy; there also seems to be growing willingness to disregard it, or trade it for other important interests" (p. 335). They point to something that has only risen more prominently since September 11: "in the wake of a series of heinous terrorist attacks, privacy may have to give way to the FBI's request for a broad range of new surveillance measures" (p. 336). Alderman and Kennedy express the position that many other people would probably agree with (according to surveys), that is that privacy is desirable but sometimes other things outweigh it (such as "the war on terrorism"). One Canadian survey, for instance, found that 58 percent of respondents believed that terrorism outweighed the protection of individual's rights and the due process of law.[12] When the question is asked about effects on the respondent this number usually declines (i.e., people are not so willing to have their privacy outweighed). People don't seem to mind giving up privacy in the abstract if it also attains another social good – safety – but they do seem to mind giving up their own privacy. Another way to put this is that people are willing for others to lose their privacy (because they are law-breakers or dangerous) but are very unwilling to lose their own privacy. On this view, privacy must yield to the state's own right to pursue its goals. But since most people see themselves as law-abiders such a loss of one's own privacy is unlikely.

At-risk behavior, trangression of "norms", or membership of a problematized group will result in an "acceptable-to-most" loss of privacy (according to these types of surveys). Institutions and government can trade on these opinions (as well as help engender them) by making dichot-omies such as the ones discussed in this chapter: normal/abnormal, risky/safe, at-risk resource/safe resource, dangerous/normal. Thus government can say "you have nothing to fear if you've done nothing wrong". Inversely, if you have done something "wrong", you have some-thing to fear until you do what's "right" (this is the argument the Bush administration uses with regard to Iraq). These are both very powerful statements because they establish the issue as one of normalization. In this chapter I have argued that the issue is not what's "right" or "wrong" as much as it is one of examining and resisting the effects of normaliza-tion "behind" the establishment of right and wrong, normal and abnor-mal as such. Right and wrong are thus derivative of rationality.

Lyon echoes the fact that privacy is thought to be "at risk from large and powerful agencies" (p. 14). However, he also calls for a better under-standing of privacy. What is privacy after all and which domain of knowl-edge shall we use to answer this question? The legal domain? The political-liberal "public" versus "private" sphere (state/private individ-ual)? A philosophic-cultural distinction between public discourse (e.g., religion) and private life (Rorty, 1999)? Lyon notes that these are all prob-lematic because they have all undergone a blurring, which "throws into radical doubt the usefulness of 'privacy' as a concept that can cope soci-ologically (let alone legislatively!) with the challenge of electronic surveil-lance" (Lyon, 1994, p. 16).

Stalder (2002) makes the same point a little more forcefully. He points out that privacy is "notoriously vague" (p. 121). One definition, estab-lished in European law, is:

> where privacy is understood as "informational self-
> determination". This, basically, means that an individual should
> be able to determine the extent to which data about her or him is
> being collected in any given context. Following this definition,
> privacy is a kind of bubble that surrounds each person, and the
> dimensions of this bubble are determined by one's ability to
> control who enters it and who doesn't. Privacy is a personal space;
> space under the exclusive control of the individual.
>
> (Stalder, 2002, p. 121)

But such a definition is unworkable because only the most ardent privacy advocate can manage all the decisions and pay attention to all the potential threats to this privacy "bubble".

Privacy then is a troubled organizing principle for any resistance to surveillance because:

1. The public–private distinction has been all but eroded
2. It is socially constructed as a component of normalization

How then might we clarify these issues, in particular the issue of the spatio-disciplinary society that is now in effect? The answer suggested in this chapter is through an outline of the structure of the rationality of carto-security. Let me now summarize how this rationality can produce risks of security.

THE RISKS OF SECURITY

I have argued in this chapter that we should, so to speak, examine the rationalities "behind" the deployment of surveillance. Why has surveillance been deemed a useful practice? What were its initial goals? We saw that historically they were implemented in order to practice government. Once "security" and "surveillance" are established as the issues there is only a very narrow range of possible statements. After all, who wants to be "against" security? Furthermore, once security is the issue, it requires an approach that constitutes people and the environment as at-risk resources. Historically statistics and statistical mapping arose in conjunction with each other to provide quantitative measures of risk, resource value, and territorial distribution. The realm of digital crime-mapping, here examined under the question of how subjectivity is formed in cyberspace, was taken as our example.

In this discussion I have raised a number of issues that I believe are germane to the role of mapping and GIS in the United States. Working within a broadly Foucauldian perspective, I have especially highlighted how a *rationality of carto-security* is constructed in which geo-surveillance is deployed as a response to dangerousness, and in which people are constructed as at-risk resources subject to normalization and management. I would now like to draw a few general conclusions concerning what might be called the "risks of security".

1. The issue of security is often contrasted against issues of privacy or civil rights. The two are seen in balance, sometimes moving more toward one side (more surveillance in times of threat) and sometimes to the other (reassertion of civil rights in times of peace). Alternatively they may be seen as opposing ends of the political spectrum (the left being pro-rights and the right being pro-security). Both of these oppositions may be at the wrong level of analysis and all too easily break down – when a rationality becomes dominant all political positions may wish to associate with it.[13] It certainly seems to be the trend over

the last decade or so that privacy has been in retreat. The fact is, the battle has largely been won in favor of a surveillant rationality even before September 11, 2001. Foucault's insight is that we should understand the rationality *itself* "behind" security, geo-surveillance and rights; that is to say what justifies it and gives it its status as truth. Opposing surveillance by appealing to civil liberties is problematic because the latter are easily constructed in different ways. As Attorney General John Ashcroft argued on the anniversary of September 11: "we're not sacrificing civil liberties. We're *securing* civil liberties. That's what our defense is. The assault on civil liberties is one by the terrorists – they are the ones who don't believe in freedom" (Williams, 2002). Ashcroft is right to make this point – what he omits is that civil liberties are increased for that set of people who are "normal" in their behaviors and that that judgement of normality can be abrogated very easily. Thus rights are always partial and limited. By pinpointing instead the rationalities at stake we can more effectively broaden the set of possible ways of thinking by showing that reason has a history that can be overcome. This was precisely the project of Foucault's genealogies and it's precisely what's at stake with regard to GIS and security.

2. At what level would interventions in the discourses occur? The argument made in this chapter is that we should necessarily be intervening at the level of rationalities because mapping (GIS and cartography) are important technologies in the production of political thought. Mapping is an important source of knowledge about questions of territory and populations. Thus the question of government must necessarily include the politics of space and the manner in which maps and GIS *produce* our knowledge of geography, people and places that in turn becomes taken up by the political process. But, we may also be interested in particular technologies or institutions (crime-mapping or ESRI) primarily insofar as they reveal the rationality in operation. It is not a question of being "anti-GIS" or anti-geo-surveillance, but rather one of critically understanding what rationalities these technologies produce, and how they are deployed in policing, policy-making, and politics. However, although useful, intervention at the level of actual practices can be easily sidestepped or diverted. For example, it would appear on the face of it that such techniques as "user-controlled privacy" which have recently been suggested by the GIScience community (Bhaduri and Onsrud, 2002) offer a technique of protecting privacy in order to receive services. However, these techniques are self-control of *loss* of privacy, rather than it being taken away by other parties (e.g., through surveillance). Thus the rationality that we need to give up privacy in order to receive services is left unchallenged. Only

an enquiry into the very rationality itself is situated to enable other ways of governing.

3. Many authors writing on geo-surveillance have noted its increased breadth and scope in the information age. Yet governmental surveillance is not new, either in the sense of digital surveillance (see Burnham's classic analysis 1984) or of the old-fashioned panopticon. Certainly it is more powerful and all-encompassing, as I noted in the first point above. But I have argued that what is important is the manner in which it is part of a governmental rationality.

4. A significant risk of security is that it builds up people as resources within a rationality of risk. But some philosophers have been arguing since the early twentieth century that it is a mistake to understand people as empirical things, because to do so closes off other possibilities. Heidegger, for example, offers the following rather prescient description of this tendency:

> Such an appraisal posits the human being as something present at hand [i.e., objectively present], deposits this thing into an empty space, and appraises it according to some table of values that is attached to it externally.
>
> (Heidegger, 2000, p. 175)

Although this was written in 1935 perhaps it could equally serve as a description of digital cartography and GIS today.

5. Finally, there is *false security*. False security is the belief or understanding that the measures we undertake are increasing security when in fact they may do no good or even cause harm. Following the September 11 attacks experts on the Middle East warned that invoking a blanket rhetoric of "terrorism" was unlikely to get at the root cause of foreign animosity towards the USA (often identified as American foreign policy, and in particular US support for Israel and its military presence in Saudi Arabia). While this is certainly a danger and a risk (what economists call an opportunity cost, that is a cost incurred for doing the wrong thing) false security is, nevertheless, still security. By this I mean that the discourse of risk assessments, and human and natural resources at-risk, is still the grounds for the constitution of the subject. My analysis here is at a different level; it seeks the motivating rationality and the resulting effects "behind" security, risk-assessment, and at-risk resource. One cannot oppose security by speaking of false security, for they are both efforts to achieve the same aim: better security.

CHAPTER 7

Geographies of the Digital Divide

As we saw in Chapter 1, cyberspace is often characterized as a free-floating domain where space has finally been overcome. Although geographers have been quick to show the inadequacy of this conception, it remains to be done to show how cyberspace is materially produced. In this chapter I offer a fuller account of this subject by examining the geographies of the digital divide.

The "digital divide" was first coined in its present sense about 1996 (for example see box) but has a much longer legacy under other names which treat the causes and effects of the uneven geographic distribution of resources (environmental justice, redlining, gentrification, etc.). But now

---

A new gulf in American education: the digital divide

The *digital divide* between . . . two schools in the heart of Silicon Valley provides perhaps the most striking example anywhere in the nation of a widening gap – between children who are being prepared for lives and careers in the information age, and those who may find themselves held back . . . There have always been educational gulfs between poor and rich schools, of course, but some experts contend that computer technology accentuates the differences.

"The way computers are used in the classroom – and the way the Internet will change their use – is really a profound commentary on education", said Michael Kirst, a professor of education at Stanford University. "The Internet is a prophetic example: richer kids with access to a home computer and to the Internet can use it as a means of exploration and discovery. Poorer kids without the Internet will just use a computer, in the classroom, for drill-and-practice exercises".

(Gary Andrew Poole, *New York Times*, January 29, 1996)

---

the term has exploded on to the popular consciousness. A recent search on Lexis-Nexis, the newspaper database, showed 585 uses in the 12 months between June 2000 and June 2001. The divide was even addressed by the candidates in the Bush–Gore Presidential race in 2000. Until the recession slowed rates of giving, grants from federal, private, and non-profit agencies to bridge the divide grew during the 1990s, and it seems as if every major city in the US and Europe deployed community technology centers (CTCs) or "cybercenters" (the independent group CTCNet now lists over 450 independent cybercenters and there are many more provided by local governments). On the other hand, the *geography* of the digital divide remains under-represented and under-studied.

Although it is not usually disputed whether or not there is a digital divide, commentators have not agreed whether the divide is significant, or even whether it is narrowing or widening. There is also the question of the spatial distribution and production of cyberspace. In other words, how does cyberspace both constitute and get constituted by *access* to the modern digital economy?

SOME TERMS AND ISSUES

Since the "digital divide" is a new term for an old concept, I want to begin by defining terms. The sense that I use here is that of the *unequal access to knowledge in the information society*. This definition is therefore not based on technology, but rather our meaningful relationship to technology. Our relationship not only includes access to computers and the Internet, but also to relevant online content, to training, to jobs in the information sector, and so on. There are at least three senses of knowledge: *to know with* (access to the tools), *to know what* (access to information), and *to know how* (how to use the tools). All three of these senses constitute the digital divide and go well beyond the technological. Furthermore, the *geography* of the digital divide addresses how the relationship between this knowledge and space is uneven across multiple scales: internationally/globally; regionally and between urban–suburban–rural. As Harvey has remarked "human beings have typically produced a nested hierarchy of spatial scales within which to organize their activities and understand their world" (Harvey, 2000, p. 75). We might ask therefore in what ways the uneven production of knowledge produces different spaces? Turning the question around, in what ways does uneven spatial distribution of resources (e.g., computers, DSL, training) produce uneven knowledges? To answer these questions, I will first provide a general picture of the various geographies of the digital divide at the three geographic scales from global to local.

The history of the digital divide no doubt extends back to the first information tool, the abacus in 3000 BC.[1] In the digital realm perhaps the first ever comment about uneven distribution of resources is attributable to Thomas Watson, founder of IBM, when he said "I think there is a world market for maybe five computers" in 1943.[2] The quotation may be of doubtful accuracy, but the sentiment is familiar; presumably the United States and its allies would have received these computers and not the rest of the world!

Nowadays the issues of the digital divide are whether access to computers and the Internet is equally available across the city. Furthermore, is this access widening or narrowing? Is access determined by income or do other spatial variables also determine access? Are there areas with higher than predicted or lower than predicted access to communication technology, given their income?

One way to answer these questions is to take surveys and collect empirical data about access rates. These surveys are invaluable in revealing the state of play, as it were, about connectivity and access. Unfortunately, surveys suffer from some inherent problems. First, they can only reflect what is actually the case at the time of the survey. They cannot predict. Of course, another survey could be carried out. But surveys are expensive and cannot necessarily be done for every place. Survey results, because they are so useful for marketing purposes, are also often only available for purchase and can cost hundreds of dollars. Second, surveys are necessarily done for a particular time and place (e.g., nationally). The NTIA surveys are an example. The NTIA collect Internet connectivity rates for the country as a whole (or at the most by regions such as "the West", "the Southeast"). They miss, therefore, the local picture. If a place wishes to locate some cybercenters across the city in areas of low Internet access, the NTIA results are not directly usable. Third, surveys use different methodologies and definitions which makes them hard to compare. When we say "Internet access" what are we measuring? Do we mean access at home or at work? If we measure at home does everybody in the household have equivalent access? What is the nature of that access – what are people using the Internet for (email, surfing, online voting)?

We can suggest here therefore that a more fruitful approach might be to develop a predictive model that could be applied at different scales (local or citywide). Furthermore, once the model is "tuned" it could be applied to different cities, obviating the need for expensive and repetitive surveys. Finally, the model would allow direct comparisons to be made across time or between different cities since it is based on a coherent methodology.

One way to think about such a model is a spatial prediction "surface" of access. Such a surface would indicate the predicted Internet access rate across the city based on known relationships between socio-economic variables such as income, race, and education and Internet access. Although these variables are quite well correlated with Internet access it is unlikely that they could perfectly predict access – there will be an "X-factor" of unexplained variation. This is because Internet access data that has been measured against the socio-economic variables mentioned above are *national averages* (e.g., the NTIA reports, which constitute the best and most reliable measures of these kinds). What is missing is exactly the local picture. The local is the "X-factor". Why places might vary from the national average is a question we shall return to shortly. But first, we need to grasp the variation in the digital divide at different scales.

THE DIGITAL DIVIDE AT DIFFERENT SCALES

This analysis provides, for the first time, a comprehensive spatial account of access to the information society. Questions include what simply is the geography of access, and is that pattern historically stable? This question should include the geography of networks and bandwidths. I will show that access to cyberspace is geographically very uneven, and that it varies widely between class, race, and income.

This unevenness can be traced at several distinct spatial scales: the international, the regional, and the local. Different efforts will be needed to address the divide at these different scales. For example, local community technology centers can be very effective at the local scale, but a different strategy is needed to work on the gap between rural and urban access, or again increasing the Internet penetration in the UK might need different strategies than in Mozambique, where half of its revenues are directed toward foreign debt.

Access to what? The digital divide is not the gap in access to computers and the Internet alone. There are other, probably more important divides: differential access to training and knowledge (both for students and for teachers, i.e. peer-mentoring); differential access to relevant content (e.g. lack of online resources for the homeless but plenty of e-commerce sites); differential take-up of resources (e.g. in high schools in the USA); and differential deployment of skills, which causes distortions in the salary market and valuation of people's skills (those with knowledge of the latest software are valued above those with better geographic understandings).

These differentials are the material realities of cyberspace. They raise the difficult question of the ethics and morality of information technology more generally.

DIVIDES AND LAGS

The digital divide is typically characterized along the variables of income/poverty, race, age, ethnicity, disability, and geography. Within the United States there have now been five reports by the Department of Commerce on the digital divide, the most recent, *A Nation Online*, appeared in February 2002 (NTIA and ESA, 2002). These reports, using data collected by the Current Population Surveys (CPS) of the Census Bureau, depict a rapid take-up of access to digital technologies and information in this country. For example, more than half of all US households now own a computer, while Nielsen//Netratings and Pew Charitable Trust surveys show that Internet penetration has reached critical mass in the US, with over half the population having access to the Internet (see box).

Nevertheless, these same studies show that this access is by no means equally distributed. Large differentials exist between people of different income levels, race, disability, and geography. Although much is made of the rapid expansion of the Internet, this is a process largely confined to developed countries: about 91 percent of the world's population do *not* have Internet access (see table 7.1). Canada and the United States account for about 30.2 percent of all Internet users worldwide, but here too, access varies widely across the country. At the regional level, areas which have historically been associated with the development of the Internet and computing (the Bay area, the Pacific Northwest, Boston, and Washington DC) have the highest Internet penetration rates in the country. In the South, while it has some of the lowest overall Internet access (Newburger, 1997), both Atlanta and Orlando are well connected and

---

Access differentials

The fifty percent mark in the United States was reached in September 2000 (according to Nielsen//Netratings, 2000). As might be expected, however, the same company shows a wide geographic disparity between different American cities; figures for March 2001 show a peak of about 69 percent penetration in San Francisco, Seattle, and Portland, to 53 percent in Houston (among those sampled). Atlanta was eleventh, with about 61 percent penetration. Obviously these figures are likely to change significantly between time of writing and publication, but the message is stable; first that there is a majority access in most major markets, and second that there remains significant geographic disparity in different regions of the country. A third conclusion will be addressed below, i.e., even wider range of access *within* these markets.

Table 7.1　The digital divide by region, late 2002.*

	Percent access to Internet 2002	Total persons (m) 2002	CAGR (%) 1997–99	PPP 2001 (US $)
USA and Canada	55.9	182.67	63.6	31,370
Europe (EMU)	32.9	190.91	46.2	24,180
Asia/Pacific	10.6	187.24	48.8	4,040
Middle East	8.2	5.12	n.a.	5,230
Latin America	5.7	33.35	111.2	7,070
Africa	1.0	6.31	30.9	1,620
Global	9.7	605.6	59.5	7,570

* All figures approximate and represent adults and children with access to the Internet.
Figures can vary widely within regions. CAGR = compound annual growth rate of
hosts per region. PPP = purchasing power parity.

Source: NUA, 2002; World Bank (2002).

have higher penetration rates than much bigger cities (for example New
York, Los Angeles, Houston, Chicago).

Finally, we may detect differentials at the local level (county and
below). According to a survey done by the *Atlanta Journal Constitution* in
June 2000, households in the Atlanta metro area may vary in computer
ownership by as much as 30 percent (54 percent in Clayton county to 84
percent in Gwinnett). By contrast, data from a UGA study indicate that
TANF (welfare) recipients across Georgia have a 15.8 percent computer
ownership rate, with TANF recipients in the suburbs outpacing those in
urban or rural areas (Larrison et al., 2000).

Variation also exists along other dimensions. In the case of African-
Americans as a whole the differential in Internet access is as much as 18
percent (23.5 percent vs. 41.5 percent access rate nationally, summer
2000). Furthermore, figures show that at least in this respect, the differential is *widening* rather than narrowing. According to the Commerce
Department, the gap between African-American and national access
rates has widened by 3 percent in two years. Critically, these differentials
cannot be entirely accounted for by income or education. When Black
households are normalized for income and education and their Internet
access rate is estimated, these two factors account for only about one half
of the actual differences. This finding is consistent with developments in
the economy as a whole: the Census Bureau reported in September 2002
that the proportion of people living in poverty in the US increased to 11.7
percent from the previous year, which is about 33 *million* people.

Why are these differentials important? For that half of Americans
which have access to the Internet it is clear that it is used in a wide variety

of ways. These include business transactions, job searches, online voting, information searches and retrieval, entertainment, and educational advancement. For the other half of America, those who are digitally divorced, it is equally clear that they are increasingly disadvantaged. In some instances there have been reports of "cyber redlining" by companies in terms of where broadband is first installed or where some online companies are prepared to deliver goods. Even without active discrimination those without access are relatively disadvantaged in using information and knowledge which is available to others.[3] The divide therefore encompasses not only access to computers, but the knowledge and information they distribute, and familiarity with their usage which is important for success in the job market.

It is important to situate what is happening in Atlanta within wider regional and global divisions. Scales are interdependent, and in a globalized economy we have to be aware of the local effects (and vice versa). Moreover, we might expect that the distribution of digital information at the national level is extremely uneven. Finally, at the international scale we can also detect marked digital divides (see Table 7.1).[4]

Table 7.1 reveals how regions with lower levels of access on the whole are also growing more slowly than regions with higher levels of access. Canada and the USA are still the predominant center of the Internet, both in terms of numbers and as a percentage of all online. However, the global share of users in these two countries has now fallen to about 30 percent, from 100 percent in the early 1990s. So the Internet is itself getting less concentrated. *But it is not flowing out evenly.* Vast areas of the world, especially in Africa, the Middle East (where there are cultural and religious barriers to adoption), and Asia, are effectively not online. As Figure 7.1 shows, the Internet is still concentrated in the US, Europe, and Australia and Japan. Only four countries in 2000 had over 50 percent access to the Internet out of about 175 for which there is data.[5] During the Internet boom of the 1990s, it was easy to imagine that everywhere would soon be connected. Post-2000 and the bursting of the dot.com bubble, this seems much less likely to happen any time soon.

Even within these countries it is important to note that access may only be available in the capital. Although most countries in the world now have Internet access, this is as misleading as saying that 95 percent of US schools are on the Internet. We must ask what is the quality of this connection and where does it go? In Africa, for example, the best connected country (South Africa) has only 7 percent of its population online (in December 2001), and there is a continent-wide average of just 1 percent (see Table 7.1). By the beginning of the twenty-first century only three out of 1,000 people even had a computer in Africa, with perhaps up to a

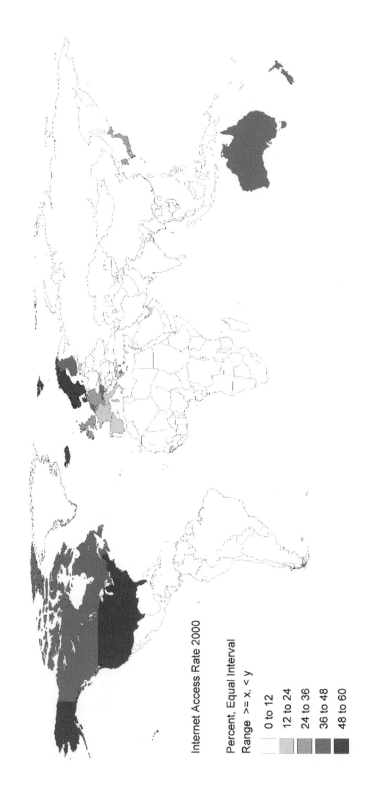

Figure 7.1 Internet access rates for 2000 by country.

Source: NUA (2002). Map by author.

The host dilemma

The most commonly (ab)used indicator to compare Internet development between countries is the number of host computers. The best known survey of Internet hosts is carried out twice a year by Network Wizards for the Internet Software Consortium. Some national Internet network administrators compile data on the number of hosts in their countries. The latest available figure from Network Wizards is 162 million hosts worldwide (Internet Software Consortium, 2002).

Network Wizards uses the following definition of hosts: "A host is a domain name that has an IP address (A) record associated with it. This would be any computer system connected to the Internet (via full or part-time, direct or dialup connections), i.e. nw.com, www.nw.com". While hosts might be a useful infrastructure indicator of the number of computers in a nation that are connected to the Internet, it is a poor indicator of accessibility since it does not measure the number of users . . . A major drawback with hosts is that they are assumed to be located in the country shown by their two-letter ISO country code Top Level Domain (ccTLD) (e.g., .nl for Netherlands). However, "There is not necessarily any correlation between a host's domain name and where it is actually located. A host with a .NL domain name could easily be located in the U.S. or any other country. In addition, hosts under domains EDU/ORG/NET/COM/INT could be located anywhere. There is no way to determine where a host is without asking its administrator" (Minges, 2000). This is a major shortcoming and results in misleading interpretation of the data.

million dialup subscribers, two-thirds of them in South Africa (Akst and Jensen, 2001). According to the United Nations *Human Development Report* (UNDP, 2001) sub-Saharan Africa has 0.6 Internet hosts per 1,000 people, compared to 179.1 for the United States and a global average of 15.1 (a host can connect more than one person).

In other words, the information connectivity of the United States is about 300 times that of sub-Saharan Africa as measured by number of hosts. Although there are problems with using hosts to estimate number of Internet *users* (see box) per country it can still give a relatively consistent picture as long as it is supplemented by other data measures (such as number of computers or telephones per capita), while the geographic distortions mainly affect small islands with fortuitous ISO codes (Tonga: .to, Tuvalu: .tv) which sell them to non-residents for memorable URLs (e.g., www.surf.to/mysite; www.whatson.tv).

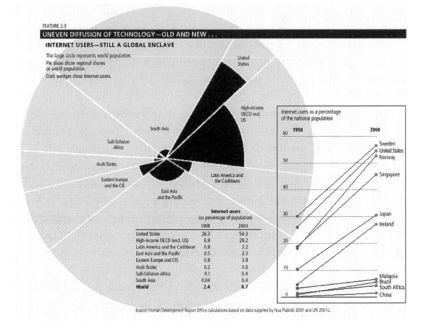

Figure 7.2   Internet users proportions for regions of the world.

From *Human Development Report* 2001 by United Nations Development Programme, © 2001 by the United Nations Development Programme. Used by permission of Oxford University Press, Inc.

Nevertheless, in Figure 7.1 I have chosen to use the latest cheaply available user access *rates*, not hosts per capita, as compiled by NUA. This map shows the global distribution of access density as a percentage of the country's total population. Only a handful of countries around the world have even moderate connectivity rates (e.g., above 30 percent), while most of the world's population finds itself with only a blank space on the map. As we know from other figures (e.g., cell phone ownership), the countries with the best connectivity to the information economy are Scandinavian. Parts of Western Europe (Spain, France, even the United Kingdom) have surprisingly low access rates, and we must ask what factors are causing this dampening effect: government policy and user perception of the technology is usually cited.

Because the map's boundaries are proportional to area, not population, small heavily populated countries can be downplayed. An alternative method is to use a "cartogram" where the size of the base units on the "map" is proportional to the size of the data. This method was employed by the UNDP (2001) as a large circle with portions or slices taking up the proportion of Internet users in each region (see Figure 7.2). The figure

The wiring of a continent

Critics don't deter Internet visionary

Quaynor has to buy more overseas bandwidth than he'd like, because little interesting Internet data is stored on Ghanaian computers. Most NCS customers want to search for information at California-based Yahoo, or send an e-mail message using Microsoft Corp.'s Hotmail service, based in Washington state. These two sites account for half of all NCS traffic.

Meanwhile, many Ghanaian firms host their Web sites in the United States or Europe, where Internet connections are faster. "That's freeloading", grumbled Quaynor. "They're not training people, so the work is being sent to the US. They're not helping build infrastructure".

So Quaynor pressures its customers to change their ways, using tactics that would outrage American Internet users. NCS monitors the traffic of its major accounts. If Quaynor feels that too much is going to non-Ghanaian sites, he'll jack up the customer's bill, or even adjust the network to reduce that user's allotment of bandwidth.

If that sounds intrusive, consider this: Quaynor also thinks Internet users in the United States, Europe, and Asia should cover part of the cost of wiring remote regions of the world, such as Ghana. He points to the international telephone market. When an American calls Ghana, he has to pay a fee to the Ghanaian phone company. But if that same American sends e-mail to one of Quaynor's customers, NCS doesn't get a dime, even though the company is paying over $1,000 a day for its Internet connection.

It's the digital divide at its worst, said Quaynor. "The de facto policy is, screw you", he complained. "It's your problem to come into the Internet. So you have to pay the whole cost to come". Quaynor's solution: a special fee to be imposed on Internet users in affluent lands to be spent on faster, cheaper network links in Third World countries.

Hiawatha Bray (*Boston Globe*, July 24, 2001)

shows that the United States has done the best at filling its "slot" but that Asia, Latin America, and Africa have barely penetrated at all. Most of the circle is empty.

Ghana makes a particularly illustrative case of the challenges that lie ahead at the international scale. Although it is a country which established one of the first gateways between the continent and the rest of the Internet (by Network Computer Systems – NCS – in the mid-1990s: see

Figure 7.3   Access to the Internet outside the primary city is still extremely rare in Africa.

Source: Mike Jensen. Used with permission.

box), by the end of 2000 there were still only four ISPs and Points of Presence (POPs) in seven cities, according to Mike Jensen, a leading researcher on African connectivity.[6] Nationwide, there are about 1.1 Internet hosts per 1,000 people which compares favorably with the rest of sub-Saharan Africa, although is still extremely low globally. However, Ghana still faces an undeveloped telephone network, with only about 70,000 lines, of which 50,000 are in the capital city, Accra. Ghana's telecom agency, Ghana Telecom, was privatized in 1995.

Figure 7.3, compiled by Jensen, shows that relatively few African countries have Internet access beyond the primary or capital city. In Ghana, several ISPs offer secondary dial-up, with access available in seven cities (Accra, Kumasi, Takoradi, Ho, Obuasi, Tamale, and Tarkwa). Africa Online, one of a few international ISPs in Africa, with operations in eight countries, offers a franchised "E-Touch" program through cybercafes in Ghana and elsewhere. Numbers are hard to estimate, but there may be as many as 150 access points in Accra, with about 480 in total across the continent. From this it serves about 130,000 customers (Marsh, 2001).[7] The E-Touch program works by having businesses purchase and install computers on their premises, which Africa Online then staffs (so far Barclays Africa and Innscor, an African fast-food chain, have signed on).

The case of Africa Online is a good example of the capital and intellectual interconnections between the developed and developing world. The company was cofounded by two Kenyan MIT engineering students, Karanja Gakio and Ayisi Makatiani, who took advantage of the knowledge they gained abroad to build out a small Internet company in 1994. Knowledge of the Internet at this time was scanty, and as Gakio said in a recent interview: "We had to create the market. If you put an ad in the paper [in Kenya] for Internet access, people didn't know what you meant" (Marsh, 2001). After a year the company became a Prodigy subsidiary when they hit financial trouble, eventually receiving about $10 million in investment from Prodigy. When Prodigy was itself bought out by a Mexican telecom company Makatiani bought Africa Online with help from Africa Lakes, a UK-based former colonial trading, automotive, and rubber plantation company. Africa Lakes now owns all of Africa Online (Africa Lakes, 2001). Africa Lakes is trying to shed its colonial old economy past and move into the information economy in Africa, and more recently the Middle East (by acquiring an Egyptian ISP). The company hopes these new activities will return it to profitability (it lost £1.5 million in the six months to March 2001) but obviously how successful it is will substantially affect Internet activity in Africa. The bottom line still seems to be that Internet success in Africa currently continues to occur with financial and intellectual assistance from abroad, although the Africa Online employees are perfectly capable of running the company once the money is provided (Africa Online recently returned a very small profit based on its subscribers and E-Touch users).

WEALTH AND CONNECTIVITY

What is not revealed as well, however, is the relationship between the wealth of a country and access. Given that income is usually held to be the primary predictor of access, and to a lesser extent education, is it possible to detect any variance between "standard of living" and access? In order to answer this question we can derive a straightforward scattergraph matching standard of living with Internet access (see Figure 7.4).

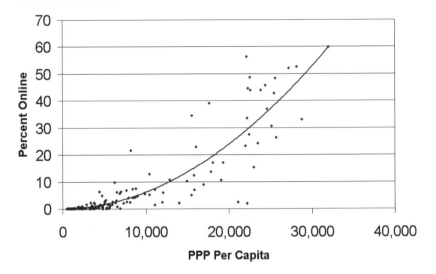

Figure 7.4   The relationship between income and Internet connectivity is non-linear. In this scatterplot, income is measured through purchasing power parity (PPP) and connectivity through percent online.

Sources: NUA (2002), World Bank (2002).

As this graph shows, income accounts for about three-quarters of the variability in Internet access, but the relationship is non-linear. In other words, adding more income ceases to have an effect on connectivity at a certain threshold level (about $21,000 in PPP international dollars; all the countries with incomes greater than this are in Europe or North America, except for Hong Kong, Singapore, and French Polynesia, i.e., former colonies). We can also identify two separable groups or clusters of countries, those with higher incomes and connectivity (though more similar in income than connectivity) and those with effectively non-access (under 10 percent and even under 5 percent access). The graph reveals an R^2 of 76 percent, which although reasonable, does not tell the whole story. That is to say, another quarter of the variability is not accounted for by income. Candidate variables which could be tested include attitudes to the Internet (perceiving it as irrelevant due to a historical lack of similar technology in the country or a lack of "killer apps"), lack of physical or cyber-infrastructure to provide access, and some related variables such as literacy, educational attainment rates, and the HIV crisis in Africa. Many countries in the developing world may be ready and willing to implement Internet technologies, but are struggling under a heavy load of foreign debt (the Heavily Indebted Poor Countries – HIPC).[8]

FROM THE GLOBAL TO THE REGIONAL: ATLANTA IN CONTEXT

If we use the definition of digital divide provided above, it is apparent that the impact is much larger than the money value of computers per se. As the Economic Policy Institute (EPI) points out "access to computers may have an important impact on a household's ability to participate in the 'new economy'" (Mishel, Berstein, and Schmitt, 2001, p. 271). According to figures collected by the US Census Bureau as well as the NTIA, home computer ownership levels in the US have been rising quickly since the early 1980s and are now above 50 percent on average (Kominski and Newburger, 1999; NTIA and ESA, 2000, p. 1). What is apparent is that this growth is taking place unevenly, so that again it may be better to speak of constant "lags" with reference to any particular technology rather than binary "divides".

For example, differentials in home computer ownership are detectable by education and by race. In both these cases there is an ownership gap. The pattern of ownership is one where there was not much difference by race or income in computer ownership in 1984 (when very few people owned computers) but that Blacks and Hispanics fell further behind the national average by the late 1990s. The technology was adopted faster among Asians and whites and the wealthy, slowest among Blacks and Hispanics. The Economic Policy Institute comments that "the patterns of computer ownership in 1998 and trends over the 1990s appear to mirror inequalities that pre-dated the era of the personal computer" (Mishel, Berstein, and Schmitt, 2001, p. 274). And although American productivity has risen substantially in the last thirty years, in part driven by investment in technology, wages have not kept pace – and wages have their own differentials by race and income. In other words, for the EPI, technology is not an "enabler" of advancement for families as, for example, the G8 digital group argue (see DOI, 2001). Or more precisely, enabling effects reinforce inequalities already in place (the rich get richer, while the poor fall further behind – see box).

In addition to race, education, and income, however, we should also look at spatial differentials. The NTIA reports data that disaggregates Internet access by urban, suburban, and rural, using US Census Bureau definitions. The 2000 report (NTIA, 2000) noted that rural areas have been catching up, if not overtaking cities. Cities have Internet access rates below the national average in America. This city lag is quite disturbing because so many people live in cities, but also because, at least in America, people who live in cities tend to be disproportionately minority and poor. In the case of the City of Atlanta for example, it is 61.4 percent black,

IT and Policy

At an IT-infrastructure workshop I attended in 2001, designed to bring together engineers and social scientists this very approach was made manifest. In response to my remarks it was observed by one participant that technology may be unequal, but at least it has raised standards for all (as if a salary increase of $75,000 for an executive was somehow equivalent to a rise in the minimum wage of 75 cents an hour!). However, it is also factually suspect. According to figures from the Economic Policy Institute, although American wages have been rising since 1997, this has not been enough to erase the decline in the previous years. In 1973 for example, the poorest 10 percent of workers were making $6.30 an hour in 1999 dollars, whereas today they earn $6.05 an hour. Wages in the top 50 percent are all higher in 1999 than 1973. Furthermore, hidden but necessary costs such as health insurance are being increasingly passed on to the worker – where in 1979 80 percent of workers received employer-sponsored health insurance, by 1998 this figure was 75 percent. The other important factor is affordable housing, which has been shrinking for some time.

compared to a Georgia state average of 28.7 percent, or the Atlanta MSA as a whole (i.e., including all suburbs) which is 28.9 percent black (US Census 2000 figures).

Figures also indicate that within cities we can find further differentials. In research performed by the Welfare Reform Evaluation Project at the University of Georgia the rate of home computer ownership was examined for TANF (welfare) recipients. As might be expected, overall computer ownership was low (15.8 percent). Another way of putting this is that the Georgia TANF rate is the same as the national rate ten years ago (Larrison et al., forthcoming). This study sampled the Georgia welfare caseload with a questionnaire which included a question on home computer ownership. The 15.8 percent who owned a computer were significantly more better off than those without (mean of $1,716.76 vs. $936.76, t = 5.965, p < 0.0001). The least likely to own computers were black welfare recipients living in the city, versus the suburbs or two categories of rurality. This means again that those who are worst off are concentrated within cities, poor or (not necessarily the same thing) minority.

Obviously a city of 416,000 people (in the case of Atlanta) will show yet further differentials between the downtown "Zone of Empowerment" to the wealthy condos of Midtown and Buckhead. But it is already clear from the discussion so far that access to the information economy in the United States is multi-faceted and complex, and spatial.

Table 7.2 shows the baseline economic data for various spatial units in Georgia and for the United States as a whole, with comparison to three other cities associated with the information economy, one in the South (Austin, Texas), one in the Northeast metropolitan corridor (Boston, Massachusetts) and one in the heart of the Internet industry (San Francisco/San Jose, California).

The Atlanta Regional Commission (ARC), which is charged with planning for the region, adopts a more usual ten-county metropolitan area (see Figure 7.5).[9] However, data are not collected by the Census Bureau at this scale, although they do provide some limited data for the City of Atlanta, reproduced below.

Some of the difficulties of this data should be made apparent at the outset. The Census Bureau definition of the Atlanta metropolitan statistical area (MSA) now consists of 20 counties, which, as is typical of MSAs, covers a rather large geographic area with a range of population densities from urban to rural (see Figure 7.5).

Employment in the information economy as a whole reached 100,000 in Georgia during the late 1990s, about half the size of that of Texas and less than a quarter of the California information economy, but comparable to that of Massachusetts. This category (NAICS 51) is quite broad and includes the motion picture and sound recording industry, as well as publishing and telecommunications. It is no surprise therefore that California has such a large information economy. To fill out the picture a little therefore I have provided contrasts with a specific subsector, that of online information services (NAICS 514191), which the Census Bureau defines as follows: "Internet access providers, Internet service providers, and similar establishments primarily engaged in providing direct access through telecommunications networks to computer-held information compiled or published by others" (US Census Bureau, 1997, n.p.). This means Internet Service Providers (ISPs) such as AOL and Mindspring, but not dot-coms such as Amazon or Yahoo!. In Georgia in 1997 this sector was worth about $93 million – compared to that of California of nearly $1.5 billion it is obvious that Georgia's Internet industry was still fairly small. It must be remembered, however, that this data only includes ISPs, many of which still retain their centers in California, and that we are also comparing states of very different sizes.

At the metropolitan level, which is more comparable than states, Atlanta appears as a leading provider of Internet services in the South, much larger than that of Austin, but still nowhere the size of San Francisco or even Boston. The census recorded only about 50,000 people

Table 7.2 Economic statistics for the information economy for Georgia and selected areas, 1997[a]

	Establishments, NAICS Sector 51, number	Receipts, NAICS sector 51 ($000)	Employees, NAICS sector 51, number	Establishments, Online Information Services	Receipts, Online Information Services ($000)	Employees, Online Information Services
Atlanta, City	–	–	–	18	56,296	250–499[c]
Atlanta, MSA	–	–	–	75	b	500–999[c]
Georgia	3,163	18,939,188	100,656	97	93,013	881
Austin, TX, MSA[d]	–	–	–	46	21,107	225
Texas	7,520	40,363,181	210,654	274	258,600	2,378
Boston	–	–	–	160	b	2,500–4,999[c]
Mass.	3,282	20,548,868	113,698	159	427,181	2,738
SF/San Jose CMSA	–	–	–	296	1,064,121	5,009
California	16,302	108,719,084	450,511	722	1,404,610	9,822
United States	114,475	623,213,854	3,066,167	4,165	8,042,568	49,935

[a] Census Bureau definitions are used for areas and places. NAICS code 51 covers "Information" as a whole. "Online Information Services" comprise code 514191.
[b] Withheld by Census Bureau for confidentiality reasons.
[c] Generalized by Census Bureau.
[d] Data for Austin MSA are only available for "Information services" which are mostly but not exclusively online.

Source: US Census Bureau.

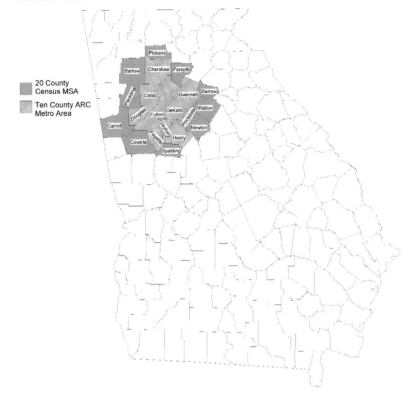

Figure 7.5 The Atlanta MSA consists of twenty counties, while its planning commission covers only ten.

Source: Map by author.

working in online information services in the whole country, and there were less than 1,000 in the Atlanta MSA. This could not therefore be described as a very significant part of the economy at that time.

That picture was not destined to last long. Just two years later Atlanta was ranked the number 3 (behind Boston and Chicago) information technology job market by *Computerworld Magazine*. Atlanta is growing fast; the workforce grew by 43.6 percent in the last decade and is predicted to be a leading center for job growth through 2025, adding some 1.8 million jobs. In 2000 there were about 120,000 companies in the area.

This growth in Atlanta was fed by a tremendous round of venture capital investment, some $2 billion for Georgia in 2000 alone, according to PricewaterhouseCoopers (cited by Davidson, 2001). The vast majority of this was in Atlanta, and about 31 percent, or $600 million, went to dot-coms in Atlanta alone. Of course, this round of investment is over for the moment.

Indeed the Atlanta area was completely omitted from a *Wired* survey of leading high-tech centers around the world (Hillner, 2000). *Wired* surveyed local leaders in government, industry, and the media to identify some 46 places around the world that make "the new digital geography". Each was rated on four measures: support from local universities, presence of established companies, new ventures, and venture capital. While the Research Triangle of North Carolina and Austin, Texas were highly rated places in the South, Atlanta was not identified as a comparable hotspot.

FROM THE REGIONAL TO THE LOCAL: ATLANTA IN DETAIL

I now wish to deal with the situation in Atlanta in more detail. Atlanta is located in Fulton and DeKalb counties in Northern Georgia with a population of some 416,000 residents within the city itself, and just over four million in the MSA. Figure 7.6 shows population density by census tract for Atlanta and surrounding areas.

As can be seen, the urban density does not fit precisely within the official city boundaries, instead stretching along Interstate 85 to the Northeast. The suburban areas sprawl extensively all around the urban area, giving Atlanta its reputation as the prototypical "sprawl city" (Bullard et al., 2000). Several surrounding cities, such as Marietta and Athens, have also reached official urban density status, and there is a string of places along I-85 and south of the city which are also in this category.

Georgia's population in the 2000 census was about 8.1 million, and it was the sixth fastest growing state during the 1990s, growing at about 26 percent over the decade – twice as fast as the United States. As a whole, Georgia is not heavily populated (its average density of 141 people per square mile makes it a "rural" state on average) but there are wide variations which must be carefully examined. In particular, the Atlanta area has shown amazing growth: on almost any figure you care to name it outperforms the rest of Georgia, the South, or the USA as a whole. The two biggest counties, not surprisingly, are Fulton and DeKalb, within which Atlanta mostly lies (about 0.8 m and 0.7 m people each) and grew at an average of 23 percent, but it is the suburban Atlanta counties which posted the biggest gains, such as Forsyth, Henry, and Gwinnett (123 percent, 103 percent, and 67 percent each).

Atlanta is both an amazingly successful technopole, attracting the headquarters and offices of many digital companies such as Mindspring/Earthlink, CNN, etc., to the extent that it has higher rates of computer

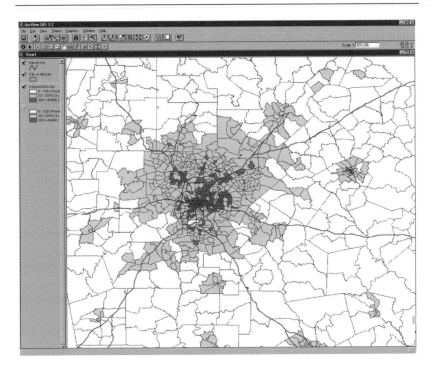

Figure 7.6 Population density in the Atlanta metropolitan region (by census tract) for 2000. Rural population densities (under 320 people per square mile) are represented by the palest shade, suburban densities (320–3,200 people per square mile) by the next darkest, and urban densities (over 3,200 people per square mile) by the darkest shade.

Data source: US Census Bureau; map by author.

ownership and Internet connectivity rates than almost any other city in the South (except for Austin, Texas), while at the same time being characterized by deep divides in income. These divides have produced a consistent geography in the region: the richer, whiter, more technological areas lie to the north, while the poorer, more minority, and lesser equipped areas lie to the south. On top of this there are local politics between the richer more outlying suburban counties and the city of Atlanta itself, for example the "regional" transportation system, MARTA, has only been built in two counties out of ten, voters in the outlying suburbs having voted against it (Bullard et al., 2000, p. 57).[10]

Let's look in more detail at the urban–suburban–rural differential mentioned above. The government reports on the divide typically analyze geographical access rates between three types of areas: urban (with population densities greater than 1,000 people per square mile, this includes some suburban areas), central cities (the largest city within a given metropolitan area as defined by the Census Bureau, such as the City of Atlanta), and rural areas (all other areas). The data show a dramatic increase in "rural" access over the past few years. In the year 2000 rural areas had an average access rate of 38.9 percent, while urban areas had an access rate of 42.3 percent (NTIA and ESA, 2000, p. 97). This represents a narrowing of the gap between rural and urban areas over the past few years. In 1997, for example, rural areas had an access rate of just 14.8 percent (NTIA and ESA, 2000, p. 5). By contrast, access rates in urban and central city locations have grown much more slowly. Indeed, in central cities access rates are now *lower* than both the national average and rural areas.

These trends are highly important. They indicate that central cities have relatively lower access rates and that they are falling behind the growth patterns of other areas (including the suburbs and rural areas). Georgia's growth was largely on the back of growth in and around Atlanta. Given that an increasing majority of people live in urban areas these differential access rates will affect many millions of Americans by providing access for some and not for others.

ADDRESSING THE DIVIDE WITH GIS

Atlanta would appear to be an excellent test case because of two competing developments: the generally lower levels of access for central cities discussed above, and the investment in the Atlanta region of communication technology (Internet, TV, and cable) which has occurred over the past decade. As a result of this investment, the Atlanta market is currently the 11th ranked city in the country in computer ownership according to the market research firm Scarborough Research (2000). Similarly, according to Nielsen//Netratings, Atlanta is ranked 11th nationwide in terms of Internet penetration (Nielsen//Netratings, 2000) with an overall Internet access rate of 54 percent and a growth rate of 18.1 percent since March 2000.

Against this backdrop the city announced in late 1999 that it had secured a large sum from the renegotiation of the cable fees operated by the then Media One, now AT&T Broadband. With this money, totaling some $8.1 million, the city planned to establish fifteen "cybercenters" to serve the needs of lower-income neighborhoods. These centers would provide classes, computing mentors, and Internet access for the public. This initiative attracted a significant amount of media attention as well as praise. What has remained unstated, however, is the locational criteria behind the allocation of the centers, apart from vague goals such as to be "within two miles of needy neighborhoods" – a fair distance to walk or travel by bus for Internet access – and to focus on the "Empowerment Zone" in Atlanta (another process imbued with local politics). Let's conclude by exploring the way GIS could help locate these centers by developing a predicted Internet access surface.

A GIS analysis of income, education and poverty in Atlanta reveals a "tale of two cities", between the haves and the have-nots. The pattern is especially noticeable because it plays out spatially between the northern neighborhoods and the southern, poorer neighborhoods. Major transportation routes such as the railroad and an Interstate help accentuate this spatial division. The Atlanta Mayor's Office of Community Technology (MOCT) was created to help address the digital divide in Atlanta and opened its first cybercenter in 2000. MOCT sees itself in the following light:

> In our knowledge-based society, information is power. How to operate a computer and use the Internet are part of the skills-set one needs today to function in society. The true value of the community technology initiative goes beyond the acquisition of technical skills; it resides in the self-discovery process that leads to self-esteem and community building.
>
> (MOCT, 2002)

MOCT does not just approach the divide as a technological problem, but rather one of community participation, power, and "cyber rights". There are now fifteen cybercenters and one mobile cyber-bus around Atlanta. A predicted Internet access surface would formally indicate where might be the best places to locate these centers. We know from the NTIA that different racial groups, incomes, and education levels have different Internet access rates. These rates are illustrated in Figure 7.7.

As both income and education increase, Internet access increases. For households with high educational and income levels, Internet access rates can be as high as 80 percent. For households in poverty and with less than a high school education, Internet access rates are as low as 5 percent (NTIA, 2000).

Taking these relationships, then, we can examine the racial, educational, and income levels for Atlanta. For each enumeration unit across Atlanta, say by census tract, we can then predict the corresponding Internet access rate as a percentage of the total population (see Table 7.3).

For example, in Tract 1 there are 88 blacks (non-Hispanic) for which we predict that 20.7 of them will have Internet access (given a national average of 23.5 percent access for blacks). There are also 3,875 whites, of which we predict 1,786.4 will have Internet access given national rates of

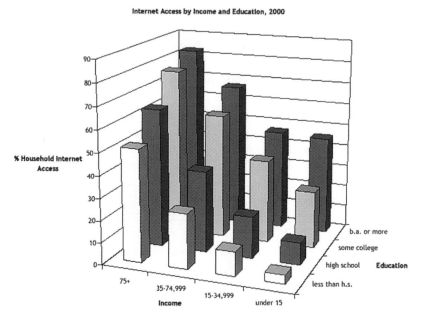

Figure 7.7   Internet access by income and education.

Source: NTIA (2000).

46.1 percent. Taking all races and ethnicities, we therefore predict that 1,862.5 people have Internet access out of a total of 4,115, or an Internet access rate of 45.3 percent for Tract 1. Taken for the city as whole, then, we might produce a map such as Figure 7.8.

The map reveals stark spatial divisions of Atlanta, especially between the north and south, and between the inner urban and outer suburban areas. These maps could then be used to inform policy decisions in locating the cybercenters and other remediation strategies.

BEYOND THE DIGITAL DIVIDE

> Once a label is on something
> It becomes an it
> Like it's no longer alive
> > *What New York Couples Fight About*, Morcheeba

As brief as our investigation of the digital divide has been we have nevertheless encountered some important considerations. The digital divide is not "purely" about access to technology, but rather to technologies

Table 7.3 Computation of predicted access by race/ethnicity.

	Tract 1		Tract 2		Tract 3	
Race	Population	Predicted household access	Population	Predicted household access	Population	Predicted household access
Black, non–Hispanic	88	20.7	2,946	692.3	349	82.0
White, non–Hispanic	3,875	1,786.4	498	229.6	850	391.9
Asian & Pacific Islander	59	33.5	24	13.6	200	113.6
Hispanic	93	21.9	72	17.0	2,641	623.3
Total predicted access		1,862.5		952.5		1,210.7
Total tract population	4,115		3,540		4,040	
Predicted access as % of population		45.3%		26.9%		30.0%

Data sources: NTIA and US Bureau of the Census (2000); table author.

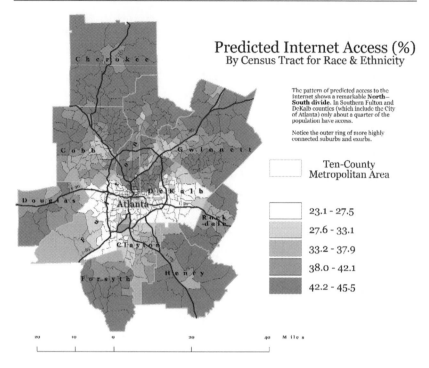

Predicted Internet Access (%)
By Census Tract for Race & Ethnicity

The pattern of predicted access to the
Internet shows a remarkable **North–
South divide**. In Southern Fulton and
DeKalb counties (which include the City
of Atlanta) only about a quarter of the
population have access.

Notice the outer ring of more highly
connected suburbs and exurbs.

Ten-County
Metropolitan Area

23.1 - 27.5
27.6 - 33.1
33.2 - 37.9
38.0 - 42.1
42.2 - 45.5

Figure 7.8   Predicted Internet access (%) by race and ethnicity for the Atlanta MSA.

that permit one to be in the world, and to encounter it in a rich and
meaningful way.

We defined the digital divide as unequal access to knowledge in the
information society. Why is access to this knowledge important? What *is*
this society? Surely the Internet is not that valuable at providing insightful analyses or critical information? In fact, is not the Internet, much like
its predecessor the television, just an entertainment mechanism that sati-
ates us at the lowest possible level? The Internet is just another leveling
down, in which any fool can express an opinion about things without
having any first-hand experience, and thus without the need to take any
responsibility (think of the strong opinions Americans have about foreign
places that they can't even locate on a map). And in any case, what with
globalization and the end of the local, the Internet is a distinct threat to
committed and essential engagement in important things (not to mention
the surveillance and disciplinary functions it seems to have).

We met these views earlier (Chapter 1) and took a stand against them
there by suggesting that the Internet, the Web, and cyberspace are not the
harbingers of disemplacement, but rather technologies that are called up
to assist in our being-in-the-world. The rather stronger version outlined

above is summarized from Dreyfus (2001), who himself is inspired by Kierkegaard's critique of the press. For Kierkegaard, the press in the mid-nineteenth century were the culprits in a leveling down by uncommitted opinion and chatter. Dreyfus argues that "[w]hat Kierkegaard envisaged as a consequence of the press's indiscriminate and uncommitted coverage is now fully realized on the World Wide Web" (2001, p. 79). For Dreyfus, on the Internet everything is leveled down and equal, and therefore has no significance at all. As he might say, we "surf" and skim along the surface without penetrating to anything essential.

But what I only stated earlier now needs to be shown more explicitly. The issue of the digital divide is not just to gain access to another entertainment medium. But we do not show this by pointing to the positive aspects of the Internet or the Web, for then cyberspace becomes something like a cost-benefit analysis of technology. The question of the digital divide is expanding the horizon of possibilities available to people, and to working as hard as we can to undercut systematic inequalities based on race, income, and geography.

Dreyfus feels that "the Net" is undermining the way we could most fully and richly live. But what is this "Net" and where is it that it could do this? Dreyfus buys his argument at the cost of objectivizing cyberspace and putting a label on it, and hence essentially killing it.

When I talk to community leaders such as Mr Kenyada in south DeKalb County, Georgia, he is not interested in a "postmodern" interplay of endlessly varying identities in chat rooms (another "leveling" idea that Dreyfus appears to derive from Turkle) but rather helping a child do some homework, and perhaps through that child getting their grandparent to learn how to use a mouse. Kenyada started a "PCs to the People" program because, as an African-American, he was struck by surveys which indicated how few African-Americans were accessing the Internet. This program refurbishes used PCs and donates them to children, youth, and seniors. He commented that many of the computers being used in Georgia to overcome the digital divide are several generations old, and added "I'd be very interested in finding out where are the biggest concentrations of antiquated computers. That kind of information would be helpful to organizations like mine that are seeking to place more Pentium-level computers in the underserved communities" (Kenyada, 2001). In a metropolitan area that has suffered decades of institutional racism, and which even today is geographically divided as Figure 7.8 indicates, what Kenyada is doing is balking at disenfranchisement, not from owning a Pentium, but from earning a living wage.

Dreyfus argues that the anonymous surfer takes no risks, for such a person "keeps open all possibilities and has no fixed identity that could be threatened by disappointment, humiliation, or loss" (Dreyfus, 2001, p. 81). Dreyfus is right to reject this free-floating notion of identity, and to speak of risk, balking, and speaking out to power. But this is exactly what Kenyada, bloggers, "parrhesiastes", and critiques of carto-rationalities are doing. When the digital divide is reduced to tracing infrastructures or mapping nodes this is exactly not the digital divide (useful though these studies are) but precisely . . . tracing infrastructures and nodes. But this is technology. The question concerning the digital divide is nothing *digital* at all.

PART IV

Conclusion

CHAPTER 8

Positivities of Power,
Possibilities of Pleasure

This book has discussed a number of issues characterized as a spatial politics of cyberspace. These problematic issues range from its uneven distribution, to the production of identity in cyberspace through geographic profiling, to a problematics at the very heart of spatial representations (maps, and particularly maps of cyberspace) themselves. It is remarkable how often these problematics emerge from the twin conjunction of space and politics.

In this final chapter I would like to offer some suggestions that may be used towards an ethics of the virtual which directly engages with space and politics. Perhaps I now more fully develop why we are necessarily concerned with space and politics, and how mapping, as a political engagement with space and cyberspace, is thus also necessarily called for. As I have stated, we are questioning the present moment, specifically the way maps constitute and are constitutive of a rationality of security, authenticity-as-identity, confessionalized truth of the landscape, and are "at a distance" from the world. "Simply" by questioning, we can begin to see how things are and outline the possibilities for change. This is the political impetus from which springs both a struggle over how best to describe the world, and how best to intervene. Nietzsche hinted in his *Untimely Meditations* why such a political impetus holds out a positivity: "that is to say, acting counter to our time and thereby acting on our time and, let us hope, for the benefit of a time to come" (Nietzsche, 1997, p. 60). In other words, when we act counter to our time we are in fact acting precisely on our time. We are thrown into a world not of our own making, but we can make it our own by grasping and taking charge of our own possibilities – what Heidegger called being authentic.

This political project is to historicize the present, which does not mean the study of something which is past: "history is not equivalent to what is past; for this is precisely what is no longer happening" (Heidegger, 2000, p. 46). Our question is rather a historicized one in order to open up possibilities in the future. By asking a question, we critique the present in order to overcome it and change it, to open up new possibilities.

In this political project that begins with a questioning, *what* do we question? For Heidegger we can only begin with what we are in our average everydayness, for this is ontically prior (see Heidegger, 1962, §4). Yet this way we are most of the time, our everydayness, is exactly what goes unquestioned. Heidegger wanted to get back to who we are as those beings for whom our being is an issue (Dasein). As we saw, Heidegger gives his answer in Division I of *Being and Time* of the who of Dasein by indicating that it is that being that is being-in-the-world. Da-sein is being *there*, but not in the Cartesian sense of a body extended in space with a location in a coordinate system, but which opens up a *world* of meaning. This is suggestive, but Heidegger does not take up that "process" of being-in-the-world that is part of our encounter with the world; namely mapping. I think this is an oversight, and that mapping is a *technē* as much as art or poetry. In some forms of art, for example music – and especially when it is not just being used for entertainment – it seems as if we can open up a whole world that lets us reflect a little bit more deeply on the meaning of our lives. Some music seems more likely to open up a world than other music – the American Top 40 doesn't on the face of it seem as likely as non-commercial music. But people find meaning in all kinds of different music. George Orwell may have mocked the proles singing the latest mass-produced pop song in *1984*, but for thousands of teenagers hearing this kind of music is a world-opening experience.

Likewise, mapping is how we are in our everydayness – if it's true that mapping (map production and map use) can be found in almost every culture and historical time period, as seems to be the case (Harley, 1987a). Thus our questioning is how mapping, as part of our average everydayness, is part of our being-in-the-world or what preliminarily has been identified as finding our place in the world.

MAPPING AS *BEFINDLICHKEIT* AND *VERLORENHEIT*

Befindlichkeit and *verlorenheit* are terms of some importance in Heidegger and I will appropriate them in a slightly odd way here (odd for Heidegger readers, anyway). Certainly Heidegger's *Being and Time* is an overlooked source of rich conceptual material that is suffused with a sense of place and space. Most notable is his delineation of the structure of being-in-the-world with which he undermines the Cartesian subject–object split, the

---

Befindlichkeit and mapping

When Heidegger explains his idea of being-in he bases it around the word "befindlichkeit". This is a Heideggerian neologism based on the German "Wie befinden Sie sich?" or how are you (literally, how do you find yourself?). In the traditional translation (Heidegger, 1962) this emerges as "state-of-mind" which has the unfortunate connotation of something mentalistic or cognitive (anathema to Heidegger). On the other hand it does suggest something like the *state* or place in which you find yourself. A more recent translation (Heidegger, 1996) is "attunement" which has the advantage of suggesting some linkage between the world and Dasein-in-the-world but loses the overtone of spatiality and finding oneself in the world. Dreyfus (1991) remarks on these translation difficulties and half-jokingly suggests "where-you're-at-ness" which may be just a bit too West Coast (his own solution is "affectedness"). But Heidegger introduces *befindlichkeit* to explicate "the existential constitution of the there" (Heidegger, 1996, p. 134/126) and neither affectedness nor attunement evoke this aspect of the there as ontologically constitutive of being-in as such. Rather, *befindlichkeit* can at least mean "finding one's place in the world" or, more succinctly, mapping.

---

interior mind contemplating the external world. For Heidegger, we are the "there" existentially (i.e., Da-seins or there-beings). Our being is constituted by being-in. What does this mean with respect to mapping (see box)?

Finding one's place in the world does not mean seeking and attaining a pre-given slot which is then occupied, for this would be to close off the horizon of possibilities. Nor does it mean the one in the sense of "the they" (this is Dreyfus' translation of *Das Man*). Rather, finding one's place in the world is something shared by all beings for whom being is an issue (Dasein). It is "our ownmost" as Heidegger might say. It accords with Heidegger's notions that we bring our own there with us, and are our own clearing:

> The being which is essentially constituted by being-in-the-world
> *is* itself always its "there" . . . [Dasein] *is* itself the clearing . . .
> [b]y its very nature, Da-sein brings its there along with it.
> (Heidegger, 1996, pp. 132–3/125)

Mapping is the "process" of making this clearing in order to bring along our being-in-the-world (our there). We engage with the world and encounter it in mapping and therefore disclose it.

What does this mean? I think that we can answer this question by comparing the traditional perspective on mapping with this more "Heideggerian" one. Traditionally, as I discuss in Chapter 5, the map is understood as a document that records a confession from the landscape about its truthful identity. Sometimes this confession is not very clear or very accurate, especially those maps from non-Western cultures or from further back in time. But starting around the Enlightenment, when rational scientific procedures were adopted such as the surveying of large areas through chain triangulation and trigonometrical surveys like the Great Trigonometric Survey (GTS) of India (Edney, 1997) then a better and better confession was coaxed out of the landscape. Eventually we've reached a stage where we are able to assess just how truthful this confession is, and to carefully sort out the truth from the lie (Monmonier, 1996). With satellite imagery that may now be as accurate as 5–7 cm in resolution, it seems that we now stand on a pinnacle of mapping achievement.

In contrast, we might respond that this accuracy is not saying anything essential about what it means to be human. Even if we go well beyond current remote sensing capabilities, say to the mapping of the human genome (a map that is actually beyond 1 : 1) have we captured the essence of the human being? What more do we essentially know about ourselves now that the genome is laid bare? Furthermore, by fixing the map as the site of a more-or-less truthful confession we are stilling the landscape, immobilizing it within a framework in which it has an identity. We saw earlier that the politics of identity was rejected by Foucault because it produces a self that is "identical to itself", that is to some sort of inner core which has to be winkled out in the confession and recorded on the map document. These identities are then "subject" to the procedures of normalization.

Heidegger's phenomenological analysis would seem to allow a more fruitful and human interpretation of mapping. A map is not a confession, although it may well be a technology, or technological practice. When we map something, we necessarily already have a vague understanding of it, but in the mapping our understanding may become more meaningful or challenged and altered. The film *The Englishman Who Went Up a Hill and Came Down a Mountain* illustrates this in a rather delightful way. Two English cartographers are mapping part of the English–Welsh border, and stay one night in a small Welsh border town. Nearby is Ffynon Gawr, the first mountain in Wales. After completing their measurements, the younger cartographer (played by Hugh Grant) has to sheepishly announce to the assembled villagers that the "mountain" is in fact a hill because it is less than 1,000 feet high. Uproar and consternation! The village orients its sense of being from being near the first mountain in Wales, which is not the highest mountain, but it is a mountain. But it's short by less than 20 feet. The "obvious" solution is that it should be raised up with earth. But is this legal or ethical? This objection gives rise to one of the greatest cartographic speeches in movie history:

> Legal? Ethical? Wh – how legal was it to say that a thousand feet is a mountain and 984 isn't, uh? Uh? Do we call a short man a boy, or a small dog a cat? No! This is a mountain, our mountain, and if it needs to be a thousand feet, then by God let's make it a thousand feet!

This "is" a mountain in its being, and if the Ordnance Survey thinks otherwise then they are mistaken. We don't call a short man a boy and we don't call a short mountain a hill. As the title says, however, the cartographer who went up a hill came down it knowing it was a mountain. His science and his accurate measurements (and his dividing practice of hill : mountain :: <1,000 feet : > 1,000 feet) were not relevant, but his mapping experience changed the way he was being-in-the-world (it didn't hurt that he got the girl as well).

Sometimes we're made aware of the need for maps because we get lost. *Verlorenheit* refers to being lost, lostness. How are we lost? We are lost "in the world", perhaps sunk into it too deeply, but so to speak in a rather superficial way. We're entangled up with trivialities and not working in particular on our own horizon of possibilities. When we're lost, the they dominate and our mode of being is inauthentic: "they authentic for-the-sake-of-which remains ungrasped, the project of one's potentiality-of-being is left to the disposal of the they" (Heidegger, 1996, p. 193/180). When we're lost, for example in the woods at dusk, meaning sinks away and we find it hard to get our bearings, our orientation to the world. Each direction seems as unlikely as the other. We're not bringing a there (a set of meaningful relations, or perhaps better *orientations*, to the world) but nowhere. We feel very anxious. It's a strong feeling; so strong in fact, that the physical state of being lost is used just as often metaphorically ("you've lost me, I don't follow what you're saying"). Having a map, or making a map, would help. It would help us find our place in the world again.

We can now see what this means for mapping. Mapping is not just using a map, but also the process of making maps as a constituent part of being-in-the-world. Mapping is also necessarily political in that our definition of the political meant the process of deciding how we decide to live in the world, and mapping is the process of finding our place in the world. Space (including cyberspace) is necessarily part of the political, and the political in turn is necessarily going to involve the question of our relation to the surrounding environment (*Umwelt*).

A critical politics understands the positivities of power and uses those positivities to push and struggle for increased human capability without necessarily falling into the trap of increased domination.

POSITIVITIES OF POWER

As we saw in Chapter 6 with reference to criminality and disciplinary spaces, mapping is a source of the production of geographic knowledge. According to Foucault there is no knowledge outside of power relations, and no power relations that do not invoke and create knowledge. Knowledge gives us the capacity to do more things, but at the same time it is used to divide, often to divide between the normal and the pathological. According to Heidegger, knowledge is an incomplete way to describe that being for whom being is an issue because it says little about our experientiality. Is there any way out of this impasse?

In Chapter 1 we considered how technologies of domination and technologies of the self come together in a contact point of governmentality. We now need to return to governmentality as the politics of ourselves, in order that we may more fully recover the positivities of power.

"In Foucault country, it always seems to be raining", says Thrift (2000). Many people see the implications of Foucault's work as negative or disempowering. Their reading of "Foucault country" is one of ever-present power which oppresses and constructs subjects in its image and in which knowledge can never shake off its disciplinary and disciplining origins. No wonder they feel that, like England, it always seems to be cloudy and wet, rather than sunny and cheerful.

But like the characterization of British weather, this is a pretty serious misreading (it rains more in Houston, Texas than in London). There are a number of reasons for this mistaken impression, at least one of which was explicitly denied in Foucault's own lifetime, which is that "there's no escape from power".

This interpretation of Foucault seems to often come from social progressives who see in his work the impossibility of escape from disciplinary power relations and who see their own work as politically committed to overturning those power relations. Nigel Thrift, quoted above, holds this position; and although he is careful to note that for Foucault power is positive, he also sees his own project as identifying a "more optimistic" view where power "is neither everywhere nor all-pervasive" (2000, p. 269). On this view Foucault voids all criteria for escaping power relations and in fact, Foucault's views are politically repressive because they offer no escape from power relations or "political quietism" (p. 269), in

Thrift's opinion. On this view, power is always bad. It's depressing, hurtful, and politically counter-productive. The trouble is, Foucault never said this, and in fact explicitly rejected it in his own lifetime (see Foucault, 1978, p. 83ff., and my Chapters 1 and 5). As he suggested himself "my point is not that everything is bad, but that everything is dangerous, which is not exactly the same as bad. If everything is danger-ous then we always have something to do" (Foucault, 1997b, p. 256).

We always have something to do. This is a positive position: not every-thing is bad but rather everything is problematizable. In Heidegger's lan-guage we can grasp the possibilities that are our ownmost and be authentically. So far from a philosophy of political quietism therefore Foucault and Heidegger are political activists.

Does this activism come at the price of the "return of the subject" in later Foucault? Some have thought so. That is, Foucault buys into a lan-guage of possibility and freedom at the cost of re-introducing the subject as something that is stable and "identical to itself" by turning to the self, an aesthetics of existence, and technologies of the self. These commenta-tors see Foucault as giving up his earlier work, which while it may have overstated the role of power at least was talking about something with political heft. But this talk of the self is a falling away into subjectivity.

We need to clear away this misunderstanding. Foucault tells us that his work was not dedicated to a theory or analysis of power: "the goal of my work during the last twenty years . . . has not been to analyze the phenom-ena of power, nor to elaborate the foundations of such an analysis. My objective, instead, has been to create a history of the different modes by which, in our culture, human beings are made subjects" (Foucault, 1983, p. 208). In fact therefore Foucault's work is a history or genealogy of sub-jectification. If, in the past, he has looked at technologies of domination in the production of subjectivity (e.g., in *Discipline and Punish*) he is now looking at the way we make ourselves into subjects, that is technologies of the self. Where these come together in a productive, positive conflict is the contact point of governmentality. He tells us more in an interview performed just five months before his death when he felt able to say what he wanted: "Power is not evil. Power is games of strategy. We all know that power is not evil!" (Foucault, 1997d, p. 298). He continues:

> I see nothing wrong in the practice of a person who, knowing
> more than others in a specific game of truth, tells those others
> what to do, teaches them, and transmits knowledge and
> techniques to them. The problem is such practices where power –
> which is not itself a bad thing – must inevitably come into play is
> knowing how to avoid the kind of domination effects where a kid

is subjected to arbitrary and unnecessary authority . . . I believe that this problem must be framed in terms of rules of law, rational techniques of government and *ethos*, practices of the self and freedom.

(Foucault, 1997d, pp. 298–9)

Presumably academics should all agree with this! So even subjectification is not "bad" in itself – we all need to be produced or we wouldn't be at all. We all need to be "thrown" in the first place. What we then do with these possibilities, indeed the way in which we try to overturn them (Nietzsche's untimeliness), is an ethics:

> For me, what must be produced is not man identical to himself, exactly as nature would have designed him or according to his essence; on the contrary, we must produce something that doesn't yet exist and about which we cannot know how and what it will be . . . it is a question rather of the destruction of what we are, of the creation of something entirely different, of a total innovation.

(Foucault, 1991, p. 121–2)

Therefore in this sense "maybe the target nowadays is not to discover what we are, but to refuse what we are" (Foucault, 1983, p. 216). Destruction of what we are . . . refusal of what we are: Foucault is close to Heidegger here with such a positivity of power in the production of subjectivity.

How do we deal with this production, how is this production of the mapper and the map-user at issue for us? It seems to me that cartographers have for the most part taken this to mean a history of accuracy in order to produce a "story of maps". Even when the project is cast as "non-progressive" (Edney, 1993) the goal is still to find a satisfactory ordering of the development of cartography as a series of events and discourses which assume a unity of terminology, resemblances, and subject matter. These histories work at the level of historical ideas and their rationales – what was the state of knowledge at this time? How did the map enterprises and the great publishing houses of Germany and the Netherlands characterize their work? Who were the important players? All of this is measured against the modern scientific conception of the map as the view from nowhere, or maps at a distance from the world, rather than constitutive of our being-in-the-world. In cartography this view from nowhere is known as the planimetric map, while in remote sensing any sense of "perspective" is corrected away into an *ortho*graphic and normalized image known as a digital orthophotoquad (DOQ). This is what Foucault warns against and knows very well how power in the sense of capacity and capability over *things* is related to power relations between individuals and groups. As we have seen with regard to the production of disciplinary spaces and dangerous individuals who pose risks, we quickly resort to a language of normality and the politics of identity which comprise a set of "dividing practices" (Foucault, 1983, p. 208) that separate the normal from the pathological.

One of the illustrations used in the French edition of *Discipline and Punish* (not reproduced in the English edition) illuminates how knowledge comes to be used for dividing practices (see Figure 8.1). The figure is used to exemplify how behaviors and practices are made to measure up against a rule(r) of correction; a standard against which deviancy (in this case the child's physical inclination away from straightness or orthopedics) can be measured, corrected toward, and managed. Even here though, power "over" others is not one of domination where all choices have been drained away, but rather one of management and government in the sense of the correct disposition of things.

Denis Wood's book *The Power of Maps* (written in 1992 in the wake of Harley's work in the late 1980s) offers a similarly illustrative example of the positivity of power. Most of the attention this book got was directed at Wood's thesis that maps have a hidden (or not so hidden) political agenda, or, as he put it, *interests*. For most of the book Wood applies a variety of techniques to uncover these interests (mostly in the vein of Barthes) and one would get the distinct impression that maps are inevitably tricksy and malicious (*Newsweek* panned the book as "conspiracies in the glove compartment?" in reference to Wood's argument that state road maps promote union-bashing and the automobile). But in the last chapter there's a seeming *volte-face*; Wood suddenly starts talking about how the "interests the map serves can be your own". This chapter always seems to be overlooked or under-appreciated by readers. But it is directly analogous to how social progressives understand power in Foucault – they see only the hidden negativities without realizing the positive productivities of these interests. Wood didn't end up arguing that we should *get out from under* the map's interests; he took the much more politically interesting route of turning those interests to other uses. And I think the best way we can understand Foucault's insistence on the disciplinary effects of power relations is to work within them to maximize opportunities and freedoms.

In this last section I want to point to one particular aspect of mapping (as constitutive of being-in-the-world) that has been overlooked. The pleasure of mapping is where space, politics, and mapping all come together in a distinctly positive manner.

Figure 8.1 "This is the rule of correction". Illustration from French edition of *Surveiller et Punir*. Original caption reads: "N. Andry. *L'orthopédie ou l'art de prévenir et de corriger dans les enfants les difformités du corps*, 1749".

POSSIBILITIES OF PLEASURE

I suggest that at least one way is through the ethics of a practice or *askēsis* which takes *pleasure* as its target, rather than desire. Such an approach was outlined very successfully in a fascinating book by Ladelle McWhorter (1997). McWhorter took her cue from Foucault's frequent interest in pleasure which forms the centerpiece of the *History of Sexuality* series. She argued that pleasure has not been politicized and is a practice which may provide a way to maximize freedoms in the context of normalization. Pleasure can be used as a counter-attack to normalization rather than "liberating" desire because, according to her, "normalizing discourses have not colonized pleasure as they have colonized desire" (p. 184). When you try to maximize freedoms through liberating desire (e.g., if gays come out) you only end up playing the rules of someone else's normalization routine. For the same reason, some argue that seeking the legal right to gay marriage should not be a political goal for gays (Warner, 1999). All it does is normalize marriage as the goal by exploiting people's feelings of shame at not being married. Why can't we just go back to the situation "before" homosexuality was invented, where there were no gays and straights, just sex acts?

McWhorter suggests a way to "refuse who we are" through a focus on a range of pleasure-practices that people are engaged in, and to destabilize essentializing sexual practices, such as the well-known "polysexuality" project attempted by the journal *Semiotext[e]*, where they wanted to map out a "cartography of desires" without reference to normality.[1] An extension of this argument is that one need not identify as gay or straight, but rather vaguely "polysexual". Instead of desire and normalization, McWhorter holds out the possibility of practices which give pleasure as the preferred political strategy: "Foucault advocates the ascetic use of pleasure as a tool for inventing a new political world" (McWhorter, 1999, p. 226).

McWhorter gives us some tantalizing glimpses of what might be done with pleasure. Her main point is that pleasure constitutes a counter-attack against normalization. A counter-attack is an active and creative opposition, rather than just a passive resistance. Therefore it already begins by refusing the terms and grounds provided by others. She writes that one effective type of opposition is a counter-attack through a "counter-memory" or counter-remembering (p. 190, pp. 206–8). The purpose of a counter-memory is to re-appropriate our lives for ourselves in a way which maximizes our capacities while breaking the link to have these capacities used for the increased dispersal of pain. Following Foucault, she notes that traditionally as people grow up their capacities increase, including the capacity to feel pain. For example, we might feel social shame for activities that as a child we carried out without shame, or indeed before even being aware that it was an "issue". Shame is an effective agent for discipline, whether it be to normalize marriage or sexual behavior. Yet increased capacities themselves (to think in new ways and to do more things) are good. When we have more capacity to do something we are in a more powerful position. McWhorter argues that therefore the key is to break the link between increased capacities and increased subjection to the discipline. She suggests that pleasure-based counterattacks or counter-memory can help break that link. If we understand the political realm as one of subjectification–resistance, or even subjectification–opposition, then it is *pain* which enables subjectification (including the threat of pain such as hunger, or juridically administered pain such as imprisonment) while *pleasure* may enable opposition and resistance. Pleasure therefore may allow us to rethink the contact point of governmentality and to rethink politics. While McWhorter identifies Foucault's study of Herculine Barbin (Foucault, 1980c) as an attempted counter-memory, here I shall explore the potential of a "counter-mapping".

As we saw in Chapter 4, Foucault argues that Greek sexuality was part of the attempt to take care of oneself (*epimeleia heautou*) by careful management of acts, gestures, and contacts, which he calls *aphrodisia* (see for example, Dreyfus and Rabinow, 1983; Foucault, 1985, p. 40). When Foucault looks back at Greek sexual practices he sees neither the licentiousness of paganism as it appears in the popular depictions of the Church, nor the repressed desires of Freudian England, but rather a set of techniques for management based on "a kind of ethics which was an aesthetics of existence" (Foucault, 1997b, p. 255).

Aristotle discusses pleasure in *Nichomachean Ethics* (Book 10). Aristotle spoke of different kinds of pleasures [*hēdonē*] such as the pleasure of learning, and pleasures which were specific to the different senses, such as sight and hearing. Aristotle understood that pleasures were not universal, that is not everybody enjoyed the same pleasures. Different people find pleasure in different activities, but in whatever they find pleasure it is so close to these activities as to be indistinguishable: "whether we choose life for the sake of pleasure or pleasure for the sake of life" (1175a 20) doesn't matter because life (as activity) and pleasure are inseparably united. Pleasure is especially good when it is "proper" to an activity, if the pleasure is not proper to an activity then it is not the best. But Aristotle makes a useful distinction in the present context, because he says that pleasures involved in activities are still more proper to them than desires [Gr. *orexis*, which has the sense of reaching or stretching out]. The reason for this is that whereas desires are at some remove from the activities by both time and nature, pleasures are close to the activities (1175b 30). They don't need to reach out for something they do not have. Desire is less connected to practice than pleasure. Interpreting this, we can say that perhaps desire is not a good way to understanding action and practices, certainly not as good as pleasure. This line of argument concurs with that of McWhorter cited above but for a somewhat different reason. It might seem a bit more existentialist (as Sartre observed, man is defined solely by his actions not by an underlying human nature). But Heidegger's response to this (Heidegger, 1993) was to say that seeing actions as only producing effects (e.g., desire–action–effect) is too utilitarian for then we only value an action by the usefulness of its effect. Action can have other purposes, he suggested (which for him as always had something to do with standing in relation to Being). Melding the arguments of McWhorter and Heidegger we can suggest that actions can have the purpose of pleasure and that the pleasure-practice pairing from Aristotle is essentially freedom.

Indeed, Foucault was interested in Greek aesthetics (practices) insofar as it constituted techniques of the governance of the self. Aesthetics is like a muscle one flexes and exercises in order to do work on oneself. And of course governance and the conduct of oneself and others is precisely the domain of politics. When Foucault talks about the practices of pleasure as a problematic, it is as a problem of ethics. In the *Use of Pleasure*, his most sustained examination of pleasure, he doesn't want to be tied down to a view of ethics as either the history of codes or of behavior in relation to those codes, but rather a third way which looks at the history of how individuals are concerned with oneself as an object, which he calls "a history of 'ethics' and 'ascetics,' understood as a history of the forms of moral subjectivation and the practices of the self that are meant to ensure it" (Foucault, 1985, p. 29). Thus we could say that our ethical project is the "politics of ourselves" (Foucault, 1999, p. 181).

In some ways, though, McWhorter, like Thrift, seems to believe that we have a route for ducking out from under power, for example by breaking the link between increased capacities and subjection. Similarly, while it is apparent that power has normalized desire, it is less apparent why pleasure could not also be normalized. Already now pleasures can be judged by a morality or normalization, e.g., ones which hurt others are bad. Maybe even concentrating on getting pleasures is narcissistic and self-involved. Nevertheless and despite these caveats there still remains "some work to be done" on how we can work on ourselves through pleasures. Practices of pleasure offer a chance to develop not an escape from politics, but a new politics in our favor: "the relationship that I think we need to have with ourselves . . . is an ethics of pleasure, of intensification of pleasure" (Foucault, 1997g, p. 131). The question is the contact point of governmentality where technologies of the self flow into and join with already existing power relations, like two tributaries coming together.

On the face of it, the pleasure of the map seems like an unlikely route to either politics or freedom maximization. But when you are looking, or being given a place, you use a map. Mapping is part of the process of creatively finding our place in the world. So when we employ pleasures of the map, the pleasure of mapping is part of the world as we relate to it. "In" cyberspace this is true to the extent that we map it to give it a place. In Dodge and Kitchin (2001) for example there is a certain pleasure in mapping cyberspace, seeing what it looks like by creating its look (see their Plates 1–8). Like all mapping, there are multiple ways of mapping cyberspace, although like all new territories it is quickly acquiring a standardized "look". Certain maps of it are becoming reproduced quite frequently. But this shouldn't prevent us from imagining other maps of it that are meaningful because they are personal and thus give pleasure.

In mapping this resistance may be found in the pleasure of mapping as a practice of finding one's place in the world. One superb example of this counter-mapping is Brian Harley's short little essay called "The map as biography" (Harley, 1987b). Harley's language in this essay is lucid and lyrical, it appeals to one's emotions as he describes a map with an intense personal meaning for him. Harley wrote this piece shortly after leaving his native England to live in America but it's not so much about his recent past as about a place he grew up and lived in for many years. Or maybe his recent move and new living place made him reflect on what other places he had lived in. Significantly it does not appear in an academic journal but rather in *The Map Collector*, in a special issue devoted to the question "my favorite map". As Gould rightly acknowledges, Harley's essay is "an intensely personal vignette of a part of his life" (Gould, 1999, p. 58) in which the map is a remembering: the "map as memory" as Gould puts it in his own piece. In fact, that issue of *The Map Collector* in the winter of 1987 is a commemorative issue devoted to remembering.

Although several writers offer their favorite maps and globes, Harley's own discussion begins by a denial, a counter-strike at the very journal which is hosting his text: "I am not a [map] collector . . ." (Harley, 1987b, p. 18). This was typical of Harley, to playfully bite the hand that fed him, but is also significant in that he clears a ground for *himself*; he reinterprets the context of his remit ("write a short piece for *The Map Collector* on your favorite map . . ."). Harley's disavowal widens the horizon of his possibilities, and questions what it means to have a relationship with maps. As he says next: "I sometimes buy and treasure maps *for very personal reasons*" (p. 18, emphasis added). Where other authors in the series chose maps for their cartographic importance historically, or their monetary exchange value, Harley immediately recasts that instrumental rationality into one of his own experiences in the most personal way. The map he chooses is an everyday one, in fact even ordinary. He "confesses" to us, but this is not a confession in the juridico-religious sense we saw earlier, but one in which he shares a set of experiential practices of pleasure: ". . . if pleasure in collecting is also aesthetic and intellectual, it is because maps can draw from the roots of our own experience" (p. 18). He might also have added, that maps constitute those roots, that maps give ground for finding our place in the world. The value of his map is not its rarity, since it is from the Ordnance Survey and printed and reprinted several times, "but in the history of how it has been used, understood, and acted upon" (p. 18) which in his case and among other things means putting it into a gilded frame on the wall, not just to be looked at, but to be toasted by the "supply of cocktails" which is conveniently nearby in his Milwaukee house.

Harley analyzes the map – "like any map" (p. 18) – in four interrelated parts, and it is interesting to see how his discussion overlaps yet radically differs from that of his colleague David Woodward. Woodward has long used a tripartite classification system to understand the history of maps: the map as artifact, as image, and as vehicle. Or in other words, the material development of the map, its look and form, and finally its social meaning (Woodward, 2001a). While these no doubt interrelate, they can be treated separately. For example, for Woodward the "map as image" leads to questions of representation and how similar (or not) the map may be from the territory it represents, a position similar if not identical to Monmonier's "lie with maps" approach.

Significantly, this scheme parallels one discussed by Heidegger in his essay concerning technology (Heidegger, 1977). Heidegger claims that today (both in the mid-1950s when the lecture was delivered, and more generally in modernity) technology is instrumental because it is charac-terized as a means to an end. This is not the true essence of technology, but it "conditions" (p. 5) how we approach it, that is instrumentality arises from technology being a means to an end, and thus where the end is a cause of that means. Heidegger then speaks of the four traditional causes of how things come about (which he locates in Aristotle). Heidegger's example turns on a silver chalice, but we can switch that to the map for our purposes.

The causes are: (1) *the causa materialis* or the material from which some-thing is made (map as artifact or product); (2) the *causa formalis* or the form and look of the material (map as image or form); (3) the *causa finalis*, or the end use to which something is put (map use); and (4) the *causa efficiens* or the means by which something is effected or brought about (the map maker). This breakdown is almost identical to Woodward's schema for the study of maps as artifact, as form/image, and as vehicle for pursuing social ends (Heidegger's fourth cause, the worker or maker, is included in Woodward's consideration of the map as artifact, including its material relations of production). Heidegger, however, is interested in questioning why these four causes and why are they similar enough to be unified? What is the character of causality, that is how things are produced and come to be? As long as we don't address these questions, Heidegger says, we will remain with an inadequate understanding of technology (including mapping) as instrumental.

Harley's own strains of analysis begin similarly, with three distinctions: the map as a physical object, as the product of its makers, and as a biography of the landscape it portrays, with a use value. But in all three he is playing off Woodward's scheme by emphasizing the *biographical*. Thus the map is a "biography of the landscape", which I think is very different from saying the map is a representation which may or may not differ from the landscape. A biography is a constructed narrative, it evokes and recalls, and helps people find their place in the world. A representation implies the need to be truthful, a will to truth as it were. This does not imply that the map as biography is necessarily untruthful – it can be and is real and true – but that a will to truth is not a hallmark of the biographical map.

To these three strains though Harley adds a fourth which does not figure in Woodward's schema, and that is the experiential relationship I have been discussing, or in Harley's words the map as personal biography: "it is a rich vein of personal history, and it gives a set of co-ordinates for the map of memory" (p. 18). This fourth aspect is of "most value" to Harley and not at all "reticent" (p. 19):

> Personal experiences and cumulative associations give to its
> austere lines and measured alphabets yet another set of unique
> meanings. Even its white spaces are crowded with thoughts as I
> whimsically reflect on its silences.
>
> (p. 20)

Although there is value to Woodward's scheme, in the final analysis the value of the map is in its pleasure of memory, because it is "an affirmation that I still belong" (p. 20). Woodward's schema also recapitulates mapping inadequately as an instrumentality. The map as memory is a practice of pleasure even when those memories are painful: "the ashes of my wife and son lie buried against a north wall of that churchyard" (p. 20). As he says "the map has become a graphic autobiography; *it restores time to memory*" (p. 20, emphasis added), by both recalling or remembering the past in the present and therefore working on oneself in the present. It is counter-memory as a practice of the self.

CONCLUSIONS

My goal in this last chapter has been to point to some political avenues that may be explored which can enrich our lives and help defeat the processes of normalization. These avenues include the practice of the map as a pleasure. It seems to me that we do not speak enough about the pleasure of mapping, and yet for many people this is how they recall their early childhood interest in place and space – the pleasures of exploring and discovery, of finding one's place in the world with a map. As well as this obvious use-value a map has a part in our current experiences as well, it is part of our world, whether for the purpose of recalling memories or for producing counter-memories. Mapping is a political project inasmuch as it is an integral part of re-politicizing in our own interests and for ourselves. One could say that because the map is as aspect of our being-in-the-world in that it "clears" the ground for politics.

   This leaves us with several implications and ways forward. First, it emphasizes counter-remembering as a crucial aspect of ethics of self-overcoming. Mapping I would argue needs to be involved in this process. Second, we need a pleasure of mapping as a counter-remembering as pointed out by Harley. Third, we need to change focus on ethics in cartography from a code to an *ethos*, a practice which is "in-place" or emplaced, where *ethos* is originally "habitat" as some writers, following a phenomenological vein, are emphasizing (e.g., Casey, 1997). Here we follow both Heidegger and Foucault in understanding that an ethics is not "a code that would tell us how to act" (Foucault, 1997g, p. 131). Since most ethical theories of computers or cyberspace are precisely guidelines, rules, and morals that tell you how to act, the ethics of cyberspace implied here are considerably different.

   My goal in this book is to examine the way that space, politics, and mapping might be productively brought together. This was done in the context of what we call cyberspace. I argued that we're often thinking about these issues at the wrong "level". Instead of opposing privacy to surveillance, for example, we should examine what gives grounds for surveillance in the first place – the governmental rationality. Similarly, we should not see cyberspace and physical space as split by a gulf that privileges physical space. But perhaps most of all I have tried to show that mapping is not a question of representing and communicating the truth of the landscape. Our human possibilities are far richer than that.

Notes

Note: All URLs were correct as of 11/2002. See References for further information.

INTRODUCTION

1. In Lotringer, 1996, pp. 416–22. Rabinow's interview "Polemics, politics, and problematizations: an interview with Michel Foucault" reproduces this paragraph (Foucault, 1997a, p. 117).

2. My usage of both "place" and "space" in this paragraph should be understood in the everyday sense, where place means local situation, as in "it's a nice place to live" and space means spatiality and spatial relations. However, given the topic of this book I also use the terms more technically, as well as to critique these common senses as not entirely adequate. In this respect, I follow Schatzki (1991) in speaking of "social space" defined as the opening up and situating in place of human life. Schatzki himself is using Heidegger's formulation of being-in-the-world.

3. See Coyne, 1998, who takes up this question using Heidegger's phenomenology to reinterpret questions of spatial proximity in cyberspace, corporeality and being-with digitally. Eldred, in an online document, also considers ontological spatiality online, locating an a priori interpretation of technology in the necessity for binary differences (ones and zeros) (Eldred, 2001). Eldred does this by reappropriating *khora* in Plato's *Timaeus* dialogue to rethink proximity online. Some of these themes are also discussed more fully in Sallis, 1999 and Derrida, 1995 (although not in the context of cyberspace).

4. This occurred in the case of Yahoo! Nazi memorabilia being offered for auction on the Web. Although France has a law against trading in Nazi materials, Web users in France could access these online auctions. A French judge threatened to fine Yahoo! if it did not control access to French Web users, which Yahoo! claimed was technologically cost-prohibitive and removed all such online auctions, even where legal outside France.

5. See also Harvey's claim that "geographical knowledges occupy a central position in all forms of political action and struggle" (Harvey, 2001, p. 233).

6. Brian Harley notes that "maps have impinged upon the life, thought, and imagination of most civilizations that are known through either archaeological or written records" (Harley, 1987a, p. 1).

7. That is, are maps non-representational objects? Do they have a status and meaning in themselves? Indeed, are they objects or practices? These are questions which have not preoccupied cartographic thinkers where textbooks and theories privilege the representational component of maps and mapping.

8. As recently as 1992 the editor of the journal *Political Geography* could write that "despite the centrality of maps to the theory and practice of political geography, we have as a subdiscipline been neglectful of our critical duties towards the products of cartography . . . *there has been no sustained effort to understand the meaning of maps for the political processes we research*" (Taylor, 1992, p. 127, emphasis added).

9. As Harvey observes, cartographic knowledges "depend heavily upon a Cartesian logic in which *res extensa* are presumed to be quite separate from the realms of mind and thought and capable of full depiction within some set of coordinates (a grid or graticule)" (Harvey, 2001, p. 220).

10. Interpretations of the recent history of Internet commercialization are contested. Certainly in the public consciousness and the popular media, as well as a number of books (e.g., Cassidy, 2002), the initial public offerings (IPOs) of many Internet start-up companies were carried out in an atmosphere of hype and fear. But this does not say much about the effects of network technology itself (just the business plans of individual companies). Nor does it address the fact that the stock market remains near its all-time high, and more than doubled between the time of Netscape's IPO and late 2001, while retail e-commerce sales alone for the 2001 first quarter were $7 billion, up 33.5 percent over the first quarter 2000 (Crampton, 2001a).

11. Foucault often distinguished between ordinary histories which concentrate on laying out and understanding events, people, and ideas, and his own project of the history of thought, constituted as problematizations (see Chapter 1). For histories of cyberspace in the former sense, see Hafner and Lyon, 1996, and Dodge and Kitchin, 2001.

12. Power is often thought of as either overwhelming and present right at the formation of knowledge or desire (so that there is no escape from it), or on the contrary power is conceived as coming from some external source and accreting to knowledge, which implies that one can be "liberated" from it. See Foucault, 1978, p. 83ff.

CHAPTER 1   BEING VIRTUALLY THERE

1. The surveillance referenced here is specifically *spatial surveillance in the era of information technologies* (Koskela, 2000). Spatial surveillance of criminal offenders is increasingly commodified, as shall be discussed in Chapter 6.

2. See Fannin et al., 2000. On virtual communities see Rheingold, 1993; Kirsch, 1995; Benko, 1997; Dodge and Kitchin, 2001. Both Hardt and Negri, and Harvey, have problematized local–global relations in the context of globalization: Hardt and Negri, 2000; Harvey, 2000, esp. pp. 54–60 and 75–7.

3. See Peterson, 1999b. Peterson, a cartographer with interests in multimedia and Internet cartography, has argued repeatedly that it is unethical to distribute maps on paper because they are so much less readily accessible than electronically distributed maps.

4. Virilio, 1997 makes some short remarks that telepresence is blurring the line between the real and the virtual (p. 45), but the comments are very brief. In general he prefers to see the Internet and telecommunication systems producing a "global delocalization of human activity" (1997, p. 82) where space will gradually yield to time.

5. The best known of these is that of "critical GIS" which sought to problematize GIS, which has now given rise to a series of books (e.g., Pickles, 1995), arguments pro and con, a listserv, and a special initiative of the multi-campus National Council on Geographic Information Analysis (NCGIA).

6. Examples of this can be found in many critical geography texts, such as Gregory's otherwise provocative engagement with what he calls the "cartography of anxiety" in his book (Gregory, 1994).

7. Yet this strategy can only be partially successful given the lack of suitable alternatives to the term (virtuality, digitality, hypertext, etc. have all been used, but suffer from the same limitations) and more importantly to *the idea of cyberspace*.

8. Foucault's word has been translated as both "problemization" and "problematization", sometimes by the same editor. In the original version of this interview, conducted by Paul Rabinow and placed in his *Reader*, the former translation is used, and in the reprint (cited here, and also edited by Rabinow) the latter. For various reasons including the fact that in this book Foucault used the French words "problématisation" and "problématique" I prefer "problematic" and "problematization".

9. I take my argument here from Unwin, 2000, especially pp. 22–3.

10. I first saw this pointed out in Derrida, 1994, p. 8.

11. Note the connection to Heidegger's "thrownness", where the question of being thrown or being-in-the world (being, thrown) is made central. It is only when something is thrown that it becomes aware (of itself) as a problem. That is, a problematization is our "project". Eldred calls this a "casting" of being (Eldred, 2001).

12. **Σωκράτης**
αγε νυν όπως, όταν τι προβάλλω σοι σοφον
περὶ τῶν μετεώρων, ευθέως υφαρπάσει
Στρεψιάδης
τί δαί; κυνηδὸν τὴν σοφίαν σιτήσομαι
Socrates
Alright. I'm going to throw you clever bits
of cosmological lore; you snap them up.
Strepsiades
I have to eat my lessons like a dog, eh?
Aristophanes, *Clouds*, lines 490–2

13. The French text says: "being offers itself *comme pouvant* (possibility) *et devant* (duty) *être pensé* (be thought)". My thanks to Camille Duchêne for providing this information.

14. The ideas of "community" and authentic places in the virtual era are in fact highly problematic. As the Yahoo! Nazi memorabilia case illuminated (Guernsey, 2001), when the local and the global collide there are multiple and complex offshoots. In this case a French judge, Jean-Jacques Gomez, ordered Yahoo! to stop providing access to Nazi memorabilia, or face a $13,000 daily fine (the local trumps the global). In turn, however, Yahoo! responded by implementing a worldwide ban across all its sites on hate-promoting material (while simultaneously challenging the order in a federal court in San Jose, California). That is, the global reproduced this local effect globally. Finally (for the moment), in November 2001, a California court determined that Yahoo! was not bound by the French decision, thus reasserting spatial differences once again. Following the events of September 11 however, it is more than likely that sites will self-censor to control hateful or terroristic speech. Thus we can at least conclude from this complicated case that the virtual and the physical are mutually intertwined.

15. My thanks to Louy Rotsos for this observation.

16. We might also note that William Gibson's famous works on cyberspace were inspired by watching kids playing video games where there seemed to him to be a complete, closed feedback loop going round through machine and player.

17. After an appeal by the government was rejected in summer 2000, the Supreme Court in May 2001 agreed to take the case. In October 2001 it referred the case back to the lower court where new arguments will be made. Thus the issue of community standards in the Internet age has not yet been directly addressed by the Supreme Court. According to the Electronic Frontier Foundation, an Internet rights organization which has filed a legal brief in the case, "only the three justices who joined the plurality opinion (Rehnquist, Thomas, and Scalia) found that geographic community standards can or should be applied to Internet speech without considering the "least tolerant community" problem. Each of the other six justices, some more strongly than others, expressed concern that the use of local community standards will cause problems for regulation of obscenity on the Internet" (Electronic Freedom Foundation, 2002).

CHAPTER 3   WHY MAPPING IS POLITICAL

1. I deviate from common practice in not capitalizing the word being because this has the effect in English of reifying the concept and making it harder to grasp.

2. For example, Harvey characterizes mapping as hegemonic: "mapping requires a map and that maps are typically totalizing, usually two-dimensional, Cartesian, and very undialectical devices with which it is possible to propound any mixture of extraordinary insights and monstrous lies" (1996, pp. 4–5). Tuathail points to power relations (re)produced through mapping: "[I]dealized maps from the center clash with the lived geographies of the margin, with the controlling cartographic visions of the former frequently inducing cultural conflict, war, and displacement" (1996, p. 2).

3. A well-known example is provided by Harley (1991, p. 16) who quotes one cartographer during the debate about the Peters projection as saying "it escapes me how politics, etc. can enter into it" (the quote is from Duane Marble).

4. Such a positive spirit of questioning the cartographic horizon can be found in Robinson and Petchenik (1976) and although they work largely within the horizon and do not engage the political their emphasis on the map as a transcendental way of knowing must be acknowledged. See, for example, the Preface.

5. The relation between Foucault's problematizations and Heidegger's ontology and especially his "equipmental breakdown" is discussed in Schwartz, 1998 and Elden, 2001a. As Polt points out, Heidegger also anticipates Kuhn's argument on paradigm shifts (Polt, 1999, p. 33, fn. 16).

6. As one referee correctly pointed out, there is also an internal disciplinary politics to protect the gatekeeping of cartographic truth from any attempts (such as Harley's) to undermine or question it.

CHAPTER 4   AUTHENTICITY AND AUTHENTICATION

1. Of course Orwell's book, *Nineteen Eighty-Four*, published in 1948, as well as a whole tradition of imaginative writing has long warned of the dangers of computers,

technology, and science, perhaps dating back to Mary Shelley's *Frankenstein, or the modern Prometheus* (1818), often identified as the first science fiction novel. However, the particular aspect I focus on here is the devolution of identity to the virtual by dint of the increasing dependence on computers to store and process personal information.

2. Note that the language of "border" "crossing into" once more discursively constructs cyberspace as an oppositional space to physical space, an objectivized space.

3. Some clarification of the terminology is warranted at this point. Although I glossed "Dasein" as "human being" it would be more accurate to add that Dasein is that human being for whom Being is precisely a question. In German the word simply means "existence" but is usually left untranslated in order to deliberately pick out its odd or uncanny appearance in English.

4. See also this comment from an interview a year later: "It's up to us to advance into a homosexual ascesis that would make us work on ourselves and invent – I do not say discover – a manner of being that is still improbable" (Foucault, 1997e, p. 163). Here Foucault clearly underlines that ways of becoming are an ascesis or practice of the self.

5. This renders the title of this book itself problematic, and even the spatially categorial usage of the word cyberspace. However, the *technē* of modern publishing requires clear titles, so my actual preferred title (*Being virtually there*) was used for Chapter 1.

6. A collection of essays called *The New Nature of Maps* appeared in 2001, but offered little that was not already published, and suffered from an introduction by a professed skeptic of Harley's project.

CHAPTER 5   COMMUNITIES IN CYBERSPACE

1. See http://www.informationweek.com/841/frontend.htm

2. See   http://www.amazon.com/exec/obidos/ASIN/0783811896/ref=ed_oe_h/102–9260382–0060907

3. See http://personalweb.about.com/library/profile/bldiarmadd3.htm. We can interpret this philosophically not as the static self being released from its depths, but dynamically as the becoming of being.

4. "I do not mean to say that liberation or such and such a form of liberation does not exist. When a colonial people tries to free itself of its colonizer, that is truly an act of liberation, in the strict sense of the word. But we also know that . . .this act of liberation is not sufficient to establish the practice of liberty that later on will be necessary for this people, this society and this individual to decide upon receivable and acceptable forms of their existence or political society".

5. Compare the sense of "ground" in Heidegger, or *Grund*, which is the world or horizon of possibilities into which we are thrown, with the additional implication that Dasein (human relation to Being) can spring forth from that ground and can change the condition of possibilities in the kynical manner described here.

CHAPTER 6   DISCIPLINARY CYBERSPACES

1. See http://www.esri.com/industries/homelandsecurity.

2. An early census is mentioned in the Bible (appropriately enough in Numbers 1, 2) and was undertaken by the Israelites as far back as the fifteenth century BC (Lyon, 1994, p. 22).

3. Political governmentality and "bio-power" were the focus of three lecture courses delivered in 1976, 1978, and 1979 ("Society must be defended", "Security, territory, and population", and "The birth of biopolitics"), the first of which appeared in French in 1997. Two of the 1976 lectures appear in *Power/Knowledge* (Gordon, 1980), and an English translation of the whole course is forthcoming. For details, see Elden, 2002.

4. The Foundation also offers advice on how to sample your child's DNA in case they are abducted.

5. Jails are temporary holding facilities for people awaiting trial or sentencing, as opposed to prisons themselves, which house offenders serving their sentences. At year-end 2001, federal prisons were operating at over 30 percent of capacity, according to the Bureau of Justice Statistics, 2002.

6. Figures for the UK are similar: between January 1999 and December 2001 a total of 44,082 people were electronically tagged and placed in Home Detention Curfew (HDC), according to figures from the Home Office. The UK also permits children as young as 12 years old to be tagged while on bail (pre-trial). Tagging was first trialed in the UK in 1995 using devices from Premier Geografix.

7. A recent job posting for a job in Boston's police department on the crime-mapping listserv lists a typical profile, remarkable for its completeness: "This position will involve conducting extensive analysis of crime and related data using Geographic Information Systems (GIS) technology and other statistical software. The Analyst will (1) perform short-term, tactical planning and analysis through cluster/hotspot identification; (2) perform long-term planning and analysis using raster-based analysis; (3) conduct crime analysis using community characteristics – land use, convenience stores, ATMs, etc.; (4) analyze temporal and spatial crime patterns in preparation for the Department's bi-monthly Crime Analysis Meetings; (5) conduct deployment and work load analysis; (6) conduct preliminary and advanced statistical analysis; (7) evaluate ongoing Department initiatives; (8) compile and analyze crime, arrests, and calls for service data; (9) identify trends and patterns in the data, generate maps, graphs, charts, and tables; (10) brief members of the Command Staff prior to the Crime Analysis Meetings on developing crime patterns; (11) write professional crime analysis bulletins relating to ongoing trends; (12) communicate with various levels of management; (13) make substantial contributions to the development and implementation of crime analysis methods; and perform related duties as required".

8. The software is known as "Rigel" and is distributed by Rossmo's company ECRI – Environmental Criminology Research Inc.

9. The grim joke was that they were pulled over for "DWB" – driving while black (Colb, 2001).

10. Christopher Priest's short story "The Watched" is an expression of such surveillance. It features tiny surveillance cameras known as scintillas that are as small and common as glitter and can penetrate everywhere in a spectacle of mutually assured observation. Also compare Philip K. Dick's tiny flying insect bots in his 1969 novel *The Galactic Pot Healer* (Dick, 1994): the AP wire service reported during summer 2002 on research at UC Berkeley on Micromechanical Flying Insects which could be used for surveillance in this manner. This research is funded by the Pentagon's Defense Advanced Research Projects Agency (DARPA) (Bridges, 2002).

11. For example, it approved $36.8 million to build a computer system to track foreign students at US universities and colleges. A GIS could very well be part of such a computer system.

12. Ipsos-Reid survey, October 2001. URL: http://www.ipsos-reid.com/media/dsp_displaypr_cdn.cfm?id_to_view=1321.

13. The Senate version of the USAPA passed 96–1.

CHAPTER 7   GEOGRAPHIES OF THE DIGITAL DIVIDE

1. This point was well captured in Stanley Kubrick's film *2001 – A Space Odyssey*, where one tribe of proto-humans gained knowledge of tools and were able to defend their territory and wipe out the other tribe. This was the first war over technology.

2. The quotation is frequently referenced. A Google search of the exact phrase yields tens of thousands of hits, often citing each other. As might be expected, however, IBM's own website gives a fuller account: "We believe the statement sometimes attributed to Thomas Watson is a misunderstanding of remarks he made at IBM's annual stockholders meeting on April 28, 1953. In referring specifically and only to the IBM 701 Electronic Data Processing Machine – which had been introduced the year before as the company's first production computer designed for scientific calculations – Thomas Watson, Jr., told stockholders that 'IBM had developed a paper plan for such a machine and took this paper plan across the country to some 20 concerns that we thought could use such a machine. I would like to tell you that the machine rents for between $12,000 and $18,000 a month, so it was not the type of thing that could be sold from place to place. But, as a result of our trip, on which we expected to get orders for five machines, we came home with orders for 18'."

3. In one of the more interesting stories, according to a study by MSNBC.com in spring 2000, Kozmo.com, an online retailer for goods, did not deliver its products to the predominantly black sections of Washington, DC. One news report on the study (not seen by this author) stated that "According to the data, nearly 350,000 of the city's 400,000 black residents live outside of the areas served by the company. But 130,000 of Washington's 170,000 white residents can get one-hour delivery from Kozmo" (see http://zdnet.com.com/2100-11-502252.html?legacy=zdnn). Kozmo denied that race was a factor, the e-tailer was not accused of doing anything illegal, and although a suit was filed by the Washington-based Equal Rights Center, it was dismissed in September 2000. The Kozmo Chief Operating Officer, Kenneth Trevathan, denied any racism, stating that, "The most important criterion for making a market decision is based on online penetration and Internet usage", which of course is the very issue under consideration here: exclusion from the knowledge economy. The ERC then reversed itself and decided to work with Kozmo after the suit was dropped to make the Internet more accessible to minorities. In the final twist, Kozmo, which like other delivery companies UrbanFetch and Webvan had been struggling financially, went out of business in April 2001.

4. This table can only be used to provide an approximate picture due to a number of variations in the data. Regional groupings obscure some significant variations. Connectivity data are obtained from surveys extrapolated to the country as a whole and may also obscure wide variations; in addition, different surveys in different parts of the world adopt different definitions of "access" (ranging from home dialup to use anywhere of the Internet within the last few weeks).

5. In order they are: the USA, Sweden, Norway, and Iceland.

6. See http://www3.wn.apc.org/africa/afrmain.htm.

7. However, the parent company, Africa Lakes Corporation plc, only reports 25,000

active subscribers and 21,200 E-Touch active users, implying a high proportion of infrequent or one-time users (see Africa Lakes, 2001).

8. According to Oxfam, 22 of 41 countries eligible for debt relief assistance were getting it at the end of 2000, but 75 percent of those were paying between 10 and 27 percent of government revenue on debt payments, and ten countries were paying more on debt than on primary education and health combined. See Oxfam, 2001.

9. The city of Atlanta predominantly occupies Fulton county with a small portion in DeKalb (see Figure 7.5).

10. MARTA is basically a City of Atlanta service, with about 45 percent of its riders living in the city, and is paid for through a 1 percent sales tax in Fulton and DeKalb counties.

CHAPTER 8   POSITIVITIES OF POWER, POSSIBILITIES OF PLEASURE

1. The editor, Sylvere Lotringer, explained it as follows:

> "Polysexuality" was an attempt to confuse the issue of sexual difference and identity by making any reference to a norm or a normality impossible. For this we created ad hoc categories – corporate sex, soft sex, liquid sex, etc. – that overlapped all moral or biological distinctions. We were taking Freud's assertion that libido has no gender literally. When I first discussed this project with the editor, François Peraldi, a heterodox Freudian psychoanalyst I had met through Félix Guattari in the early "70s, what we had in mind was rather polymorphous and anarchical. It was a political statement about sexuality, but done with humor and a lust for life. It was a way of looking outward and connecting sexuality to all the other flows that permeate society . . . The initial idea of "Polysexuality" was to map out a "cartography of desires", an erotic city. We called for practical tips and locations, unusual practices, concrete experiences.

These answers were given in an issue of "TeleSymposia" at http://www.echonyc.com/~trans/Telesymposia3/Lotringer/SLotringer.html.

References

Note: All URLs are functioning as of 11/2002. In some cases where the original URL became unavailable between date of initial access and this date, a citation to the document at the Internet Archive (http://webdev.archive.org/) has been provided. If a URL listed below is now non-functioning, the Internet Archive can often provide a copy.

Date of original publication of material in translation is given where relevant.

Aberley, D. (ed.). 1993. *Boundaries of Home: Mapping for Local Empowerment*. Gabriola Island, BC and Philadelphia, PA: New Society Publishers.

Africa Lakes Corporation. 2001. Quarterly Report 2001 [2nd quarter]. Online. http://www.africanlakes.com/pdf/q2.pdf

Akst, D. and Jensen, M. 2001. Africa Goes Online. Digital Divide Network. Online. http://www.digitaldividenetwork.org/content/stories/index.cfm?key=158

Alderman, E. and Kennedy, C. 1997. *The Right to Privacy*. New York: Vintage Books.

American Library Association (ALA). 2001. American Library Association Votes to Challenge CIPA. Press Release. Online. http://www.ala.org/cipa/ftrfnewsarticle.html

Anderson, R. H., Bikson, T. K., Law, S. A. and Mitchell, B. M. 1995. *Universal Access to E-Mail. Feasibility and Societal Implications*. Santa Monica, CA: RAND.

Andrews, J. H. 1996. What was a map? The Lexicographers reply. *Cartographica*, 33(4): 1–11.

APBNews. 2000. "The Enforcer": RoboCop of Sex Offender Pursuit. Online. http://216.239.51.100/search?q=cache:3GmvBZaMPhQC:www.apbnews.com/safetycenter/family/2000/03/28/enforcer0328_01.html

Aungles, A. and Cook, D. 1994. Information Technology and the Family: Electronic Surveillance and Home Imprisonment. *Information Technology and People*, 7: 69–80.

Barrett, C. 1999. More about Weblogs. Online. http://www.camworld.com/journal/rants/99/05/11.html

Baudrillard, J. 1994. *Simulacra and Simulation*. Trans. S. F. Glaser. Ann Arbor, MI: University of Michigan Press.

Benedikt, M. 1991. *Cyberspace, First Steps*. Cambridge, MA: MIT Press.

Benko, G. 1997. Introduction: Modernity, Postmodernity and the Social Sciences, in G. Benko and U. Strohmayer (eds), *Space and Social Theory, Interpreting Modernity and Postmodernity*. Oxford: Blackwell, pp. 1–44.

Berners-Lee, T. 1989. Information Management: A Proposal. Online. http://www.w3.org/History/1989/proposal.html

Bhaduri, A. and Onsrud, H. 2002. *User Controlled Privacy Protection in Location-Based Services*. Paper given at the GIScience 2002 Conference, Boulder, CO.

BI Inc. 2002. BI HomeGuard 200. Online. http://www.bi.com/

Biderman, A. D. 1990. The Playfair Enigma: the Development of the Statistical Representation of Statistics. *Information Design Journal*, 6: 3–25.

Billinge, M., Gregory, D. and Martin, R. (eds). 1983. *Recollections of a Revolution, Geography as a Spatial Science*. New York: St. Martin's Press.

Black, J. 1997. *Maps and Politics*. Chicago: University of Chicago Press.

Blood, R. 2002. Weblogs: A History and Perspective. In Perseus Publishing (eds), *We've Got Blog, How Weblogs are Changing Our Culture*. Cambridge, MA: Perseus Publishing, pp. 7–16.

Bloomfield, B. 2001. In the Right Place at the Right Time: Electronic Tagging and the Problems of Social Order/Disorder. *Sociological Review*, 49: 174–201.

Brand, S. 1987. *The Media Lab, Inventing the Future at MIT*. New York: Viking.

Brenner, N. and Elden, S. 2001. Henri Lefebvre in Contexts: An Introduction. *Antipode*, 33: 763–68.

Bridges, A. 2002. Tiny Flying Robots: Future Masters of Espionage, Exploration. *AP Wire*, July 27.

Brooks, P. 2000. *Troubling Confessions. Speaking Guilt in Law and Literature*. Chicago: University of Chicago Press.

Brown, D. 2001. Free Access to Medical Journals to be Given to Poor Countries. *Washington Post*, July 9, p. A12.

Bukatman, S. 1993. *Terminal Identity. The Virtual Subject in Postmodern Science Fiction*. Durham, NC: Duke University Press, 1993.

Bullard, R. D., Johnson, G. S. and Torres, A. O. (eds). 2000. *Sprawl City. Races, Politics, and Planning in Atlanta*. Washington, DC: Island Press.

Bunge, W. 1966. *Theoretical Geography*, 2nd edn. Lund, Sweden: C. W. K. Gleerup.

Bureau of Justice Statistics (BJS). 2002. *Prison and Jail Inmates at Midyear 2001*. Online. http://www.ojp.usdoj.gov/bjs/pub/pdf/pjim01.pdf

Burnham, D. 1984. *The Rise of the Computer State*. New York: Vintage Books.

Buttenfield, B. P. 1999. Usability Evaluation of Digital Libraries. *Science and Technology Libraries*, 17(3–4): 39–59.

Cairncross, F. 1997. *The Death of Distance, How the Communications Revolution will Change Our Lives*. Boston, MA: Harvard Business School Press.

Campbell, C. S. and Egbert, S. L. 1990. Cartographic Animation: Thirty Years of Scratching the Surface. *Cartographica*, 27(2): 24–46.

Canedy, D. 2001. Tampa Scans the Faces in Its Crowds for Criminals. *New York Times*, July 4.

Carr, D. B., Olsen, A. R., Courbois, J. P., Pierson, S. M. and Carr, D. A. 1998. Linked Micromap Plots: Named and Described. *Statistical Computing and Graphics Newsletter*, 9(1): 24–32.

Casey, E. S. 1997. *The Fate of Place. A Philosophical History*. Berkeley, CA: University of California Press.

Cassidy, J. 2002. *Dot.con*. New York: HarperCollins.

Catholic Encyclopedia. 1913. World Wide Web edition. Entry on "Fourth Lateran Council". Online. http://www.newadvent.org/cathen/09018a.htm.

Cerny, J. W. 1972. Use of the SYMAP Computer Mapping Program. *Journal of Geography*, 71(3): 167–74.

Chabrow, E. 2002. Every Move you Make, Every Breath you Take . . . *Information Week*, August 30.

Chrisman, N. 1988. The Risks of Software Innovation: a Case Study of the Harvard Lab. *The American Cartographer*, 15(3): 291–300.

Christensen, K. 1982. Geography as a Human Science: A Philosophic Critique of the Positivist–Humanist split. In P. Gould and G. Olsson (eds), *A Search for Common Ground*. London: Pion, pp. 37–57.

Citizen Corps. 2002. Operation TIPS Update. August 13. Online. http://www.citizencorps.gov/enews/tips02.html

Clarke, K. C. 1995. *Analytical and Computer Cartography*. 2nd edn. Englewood Cliffs, NJ: Prentice Hall.

Colb, S. F. 2001. The New Face of Racial Profiling: How Terrorism Affects the Debate. FindLaw. Online. http://writ.findlaw.com/colb/20011010.html

Cook, D., Symanzik, J., Majure, J. J. and Cressie, N. 1997. Dynamic Graphics in a GIS: More Examples Using Linked Software. *Computers and Geosciences: Special Issue on Exploratory Cartographic Visualization*, 23(4): 371–85.

Coppock J. T. and Rhind D. W. (1991). The History of GIS. In D. J. Maguire, M. F. Goodchild D. W. and Rhind (eds), *Geographical Information Systems: Principles and Applications*, Volume 1. New York: John Wiley & Sons, pp. 21–43.

Cosgrove, M. (ed.). 1999. *Mappings*. London: Reaktion Books.

Coyne, R. 1998. Cyberspace and Heidegger's Pragmatics. *Information Technology and People*, 11(4): 338–50.

Craig, W. J., Harris, T. M. and Weiner, D. (eds). 2002. *Community Participation and Geographic Information Systems*. New York: Taylor & Francis.

Crampton, J. W. 1994. Cartography's Defining Moment: The Peters Projection Controversy, 1974–1990. *Cartographica*, 31(4): 16–32.

Crampton, J. W. 1996. Bordering on Bosnia. *GeoJournal*, *39(4)*: 353–61.

Crampton, J. W. 1999a. Virtual Geographies: The Ethics of the Internet. In J. Proctor and D. Smith (eds), *Ethics in Geography, Journeys in a Moral Terrain*. New York: Routledge, pp. 72–91.

Crampton, J. W. 1999b. Online Mapping: Theoretical Context and Practical Applications. In W. Cartwright, M. Peterson and G. Gartner (eds), *Multimedia Cartography*. Berlin: Springer Verlag, pp. 291–304.

Crampton, J. W. 1999c. Development of Three-Dimensional Online Mapping Visualizations. In C. P. Keller (ed.), *Proceedings of the 19th International Cartographic Association*. Ottawa: ICA, pp. 721–8.

Crampton, J. W. 2001a. Maps as Social Constructions: Power, Visualization and Communication. *Progress in Human Geography*, 25(2): 235–52.

Crampton, J. W. 2001b. Cyber Rights and Cyber Maps. *Cartographic Perspectives*, 38: 77–80.

Crampton, J. W. 2002a. Interactivity Variables in Geographic Visualization. *Cartography and GIS*, 29(2): 85–98.

Crampton, J. W. 2002b. Cabbagetown, Atlanta: (Re)Placing Identity. *Reconstruction*, 2(3). Online. http://www.reconstruction.ws/home2.htm

Crane, G. R. (ed.). 2001. *The Perseus Project*. Online. http://www.perseus.tufts.edu

Crang, M., Crang, P. and May, J. 1999. *Virtual Geographies*. London & New York: Routledge.

Cutter, S. L., Richardson, D. and Wilbanks, T. 2002. *The Geographical Dimensions of Terrorism: Action Items and Research Priorities*. Washington, DC: Association of American Geographers.

Davidson, C. 2001. Dot-Com Blues. Exiles from Atlanta's New Economy Rethink the Dot-Com Dream. *Creative Loafing*, July 4 – July 10, pp. 29–32.

Dent, B. 2000. Brief History of Crime Mapping. In L. S. Turnbull, E. Hallisey Hendrix and B. D. Dent (eds), *Atlas of Crime, Mapping the Criminal Landscape*. Phoenix, AZ: Oryx Press.

Derrida, J. 1994. *Specters of Marx*. New York and London: Routledge.

Derrida, J. 1995. *On the Name*. Trans. T. Dutoit. Stanford, CA: Stanford University Press.

Dick, P. K. 1980. *The Golden Man*. New York: Berkley.

Dick, P. K. 1987. In D. S. Apel (ed.), *Philip K. Dick, the Dream Connection*. San Jose, CA: Permanent Press.

Dick, P. K. 1992. *The Eye of the Sibyl*, Volume 5 of the Collected Short Stories. New York: Citadel Press.

Dick, P. K. 1994. *Galactic Pot Healer*. New York: Vintage Books.

Digital Opportunity Initiative (DOI). 2001. Creating a Development Dynamic. Online. http://www.opt-init.org/framework.html

Dodge, M. and Kitchin, R. 2001. *Mapping Cyberspace*. London and New York: Routledge.

Dreyfus, D. L. 2001. *On the Internet*. London and New York: Routledge.

Dreyfus, H. L. 1991. *Being-in-the-World*. Cambridge, MA: MIT Press.

Dreyfus, H. L. and Rabinow, P. 1983. *Michel Foucault: Beyond Structuralism and Hermeneutics*, 2nd edn. Chicago: University of Chicago Press.

Driver, F. 1985. Power, Space, and the Body: A Critical Assessment of Foucault's *Discipline and Punish. Environment and Planning D: Society and Space*, 3: 425–46.

Eagleton, T. 1983. *Literary Theory: an Introduction*. Oxford: Blackwell.

Edney, M. H. 1992. J. B. Harley (1932–1991): Questioning Maps, Questioning Cartography, Questioning Cartographers. *Cartography and GIS*, 19: 175–8.

Edney, M. H. 1993. Cartography Without "Progress": Reinterpreting the Nature and Historical Development of Mapmaking. *Cartographica*, 30(2/3): 54–68.

Edney, M. H. 1996. Theory and the History of Cartography. *Imago Mundi*, 48: 185–91.

Edney, M. H. 1997. *Mapping an Empire, The Geographical Construction of British India 1765–1843*. Chicago: University of Chicago Press.

Edney, M. H. 1999. J. B. Harley's Pursuit of Specific Theories and of the Big Picture. Paper given at the 18th International Conference on the History of Cartography, Athens. 7pp.

Edwards, A. C. 1969. Walker's 1870 *Statistical Atlas* and the Development of American Cartography. Unpublished Master's Thesis, University of Wisconsin.

Egan, J. 2000. Gay, 15 and Out (in Cyberspace). *New York Times Magazine*, December 10.

Ehrenreich, B. 2001. *Nickel and Dimed. On (Not) Getting by in America*. New York: Metropolitan Books.

Elden, S. 2000. Rethinking the Polis, Implications of Heidegger's questioning of the political. *Political Geography*, 19: 407–22.

Elden, S. 2001a. *Mapping the Present. Heidegger, Foucault, and the Project of a Spatial History*. London: Continuum.

Elden, S. 2001b. The Place of Geometry: Heidegger's Mathematical Excursus on Aristotle. *Heythrop Journal*, 62: 311–28.

Elden, S. 2002. The War of Races and the Constitution of the State: Foucault's "Il faut défendre la société" and the Politics of Calculation. *Boundary 2*, 29: 125–51.

Eldred, M. 2001. Draft casting of a digital ontology. Online. http://www.webcom.com/artefact/dgtlon_e.html

Electronic Freedom Foundation (EFF). 2001. EFF Analysis of the Provisions of the USA PATRIOT Act. Online. http://www.eff.org/Privacy/Surveillance/Terrorism_militias/20011031_eff_usa_patriot_analysis.html

Electronic Freedom Foundation (EFF). 2002. Active EFF Cases and Efforts. Online. http://www.eff.org/Legal/active_legal.html

Fannin, M., Fort, S., Marley, J., Miller, J. and Wright, S. (eds). 2000. The Battle in Seattle: A Response from Local Geographers in the Midst of the WTO Ministerial Meetings. *Antipode*, 32(3): 215–21.

Federal Bureau of Investigation (FBI). 2001. *FBI Laboratory 2000*. Washington DC: US Dept. of Justice.

Foresman, T. W. (ed.). 1997. *The History of Geographic Information Systems: Perspectives from the Pioneers*. Saddle River, NJ: Prentice-Hall.

Foucault, M. 1972 [1969]. *The Archaeology of Knowledge*. New York: Pantheon Books.

Foucault, M. 1977 [1975]. *Discipline and Punish*. Trans. Alan Sheridan, 1977. New York: Vintage Books.

Foucault, M. 1978 [1976]. *The History of Sexuality An Introduction*. Volume I of *The History of Sexuality*. Trans. Robert Hurley. New York: Vintage Books.

Foucault, M. 1980a. Power and Strategies. In C. Gordon (ed.), *Power/Knowledge, Selected Interviews and Other Writings 1972–1977*. New York: Pantheon Books, pp. 134–45.

Foucault, M. 1980b [1976]. Two Lectures. In C. Gordon (ed.), *Power/Knowledge, Selected Interviews and Other Writings 1972–1977*. New York: Pantheon Books, pp. 78–108.

Foucault, M. 1980c [1978]. *Herculine Barbin, Being the Recently Discovered Memoirs of a Nineteenth-Century French Hermaphrodite*. Trans. R. McDougall. New York: Pantheon.

Foucault, M. 1980d. Truth and Power. In C. Gordon (ed.), *Power/Knowledge, Selected Interviews and Other Writings 1972–1977*. New York: Pantheon Books, pp. 109–33.

Foucault, M. 1983. The Subject and Power. In H. L. Dreyfus and P. Rabinow (eds), *Michel Foucault: Beyond Structuralism and Hermeneutics*, 2nd edn. Chicago: University of Chicago Press, pp. 208–26.

Foucault, M. 1985 [1984]. *The Use of Pleasure*. Volume II of *The History of Sexuality*. New York: Vintage Books.

Foucault, M. 1986 [1984]. *The Care of the Self*. Volume III of *The History of Sexuality*. New York: Vintage Books.

Foucault, M. 1988a [1978]. The Dangerous Individual. In L. D. Kritzman (ed.), *Politics, Philosophy, Culture: Interviews and Other Writings, 1977–1984*. Trans. Alain Baudot and Jane Couchman. New York: Routledge, pp. 125–51.

Foucault, M. 1988b [1979]. Omnes et Singulatim. In L. D. Kritzman (ed.), *Politics,*

Philosophy, Culture: Interviews and Other Writings, 1977–1984. New York: Routledge, pp. 57–85.

Foucault, M. 1988c [1984]. The Concern for Truth. In L. D. Kritzman (ed.), *Politics, Philosophy, Culture: Interviews and Other Writings, 1977–1984.* New York: Routledge, pp. 255–67. Trans. Alan Sheridan.

Foucault, M. 1991. *Remarks on Marx.* Trans. R. J. Goldstein and J. Cascaito. New York: Semiotext[e].

Foucault, M. 1997a [1984]. Polemics, Politics, and Problematizations: An Interview with Michel Foucault. In P. Rabinow (ed.), *Ethics Subjectivity and Truth. Essential Works of Foucault 1954–1984 Vol. I.* New York: New Press, pp. 111–19.

Foucault, M. 1997b [1983]. On the Genealogy of Ethics: An Overview of Work in Progress. In P. Rabinow (ed.), *Ethics Subjectivity and Truth. Essential Works of Foucault 1954–1984 Vol. I.* New York: The New Press, pp. 253–80.

Foucault, M. 1997c [1983]. Self Writing. In P. Rabinow (ed.), *Ethics Subjectivity and Truth. Essential Works of Foucault 1954–1984 Vol. I.* New York: New Press, pp. 207–22.

Foucault, M. 1997d [1984]. The Ethics of the Concern for Self as a Practice of Freedom. Trans. P. Aranov and D. McGrawth, 1997. In P. Rabinow (ed.), *Ethics Subjectivity and Truth. Essential Works of Foucault 1954–1984 Vol. I.* New York: New Press, pp. 281–301.

Foucault, M. 1997e [1982]. Sex, power, and the Politics of Identity. In P. Rabinow (ed.), *Ethics Subjectivity and Truth. Essential Works of Foucault 1954–1984 Vol. I.* New York: New Press, pp. 163–73.

Foucault, M. 1997f [1981]. Friendship as a Way of Life. In P. Rabinow (ed.), *Ethics Subjectivity and Truth. Essential Works of Foucault 1954–1984 Vol. I.* New York: New Press, pp.135–40.

Foucault, M. 1997g [1982]. Interview with Steven Riggins. In P. Rabinow (ed.), *Ethics Subjectivity and Truth. Essential Works of Foucault 1954–1984 Vol. I.* New York: New Press, pp. 121–33.

Foucault, M. 1997h [1982]. Technologies of the Self. In P. Rabinow (ed.), *Ethics Subjectivity and Truth. Essential Works of Foucault 1954–1984 Vol. I.* New York: New Press, pp. 223–51. English original.

Foucault, M. 1999 [1980]. About the Beginning of the Hermeneutics of the Self. In J. Carrette (ed.), *Religion and Culture.* New York: Routledge.

Foucault, M. 2000a [1973]. Truth and Juridical Forms. In J. D. Faubion (ed.), *Power, Essential Works of Foucault 1954–1984 Vol. III.* Trans. R. Hurley. New York: New Press, pp. 1–89.

Foucault, 2000b. Governmentality. In J. D. Faubion (ed.), *Power, Essential Works of Foucault 1954–1984 Vol. III.* New York: New Press, pp. 201–22.

Foucault, M. 2001. *Fearless Speech.* Los Angeles: Semiotext[e].

Funkhouser, H. G. 1936. A Note on a Tenth Century Graph. *Osiris,* 1: 260–2.

Funkhouser, H. G. 1938. Historical Development of the Graphical Representation of Statistical Data. *Osiris,* 3, 269–404.

Goodchild, M. 2000. Communicating Geographic Information in a Digital Age. *Annals of the Association of American Geographers,* 90, 2: 344–55.

Gordon, C. (ed.). 1980. *Power/Knowledge.* New York: Pantheon Books.

Gould, P. 1979. Geography 1957–1977: The Augean Period. *Annals of the Association of American Geographers,* 69(1): 139–51.

Gould, P. 1999. *Becoming a Geographer.* Syracuse, NY: Syracuse University Press.

Graham, S. 1998. The End of Geography or the Explosion of Place? Conceptualizing Space, Place and Information Technology. *Progress in Human Geography*, 22(2): 165–85.

Graham, S. 1999. Geographies of Surveillant Simulation. In M. Crang, P. Crang and J. May (eds), *Virtual Geographies, Bodies, Space and Relations*. London and New York: Routledge.

Gregory, D. 1994. *Geographical Imaginations*. Cambridge, MA and Oxford: Blackwell Publishers.

Grosz, E. 1995. Women, *Chora*, Dwelling. In S. Watson and K. Gibson (eds), *Postmodern Cities and Spaces*. Oxford: Blackwell Publishers.

Guelke, L. 1976. Cartographic Communication and Geographic Understanding. *Canadian Cartographer*, 13: 107–22.

Guernsey, L. 2001. Welcome to the World Wide Web. Passport, Please? *New York Times*, March 15.

Guntrum, H. 1997. *Under the Map of Germany: Nationalism and Propaganda, 1918–1945*. London and New York: Routledge.

Hacking, I. 1990. *The Taming of Chance*. Cambridge and New York: Cambridge University Press.

Hacking, I. 2002. *Historical Ontology*. Cambridge, MA: Harvard University Press.

Hafner, K. and Lyon, M. 1996. *Where Wizards Stay up Late at Night: The Origins of the Internet*. New York: Simon & Schuster.

Hankins, T. L. 1999. Blood, Dirt and Nomograms. *Isis*, 90: 50–80.

Hannah, M. G. 2000. *Governmentality and the Mastery of Territory in Nineteenth-Century America*. Cambridge: Cambridge University Press.

Hannah, M. G. 2001. Sampling and the Politics of Representation in US Census 2000. *Environment and Planning D: Society and Space*, 19: 515–34.

Haraway, D. 1991. *Simians, Cyborgs, and Women: The Reinvention of Nature*. New York: Routledge.

Hardt, M. and Negri, A. 2000. *Empire*. Cambridge, MA and Oxford: Harvard University Press.

Harley, J. B. 1987a. The Map and the Development of the History of Cartography. In J. B. Harley and D. Woodward (eds), *Cartography in Prehistoric, Ancient, and Medieval Europe and the Mediterranean*, Vol. I of *The History of Cartography*. Chicago: University of Chicago Press, pp. 1–42.

Harley, J. B. 1987b. The Map as Biography: Thoughts on Ordnance Survey Map, Six-Inch Sheet Devonshire CIX, SE, Newton Abbot. *Map Collector*, 41: 18–20.

Harley, J. B. 1988a. Maps, Knowledge and Power. In D. Cosgrove and S. Daniels (eds), *The Iconography of Landscape*. Cambridge: University of Cambridge Press, pp. 277–312.

Harley, J. B. 1988b. Secrecy and Silences: The Hidden Agenda of State Cartography in Early Modern Europe. *Imago Mundi*, 40: 111–30.

Harley, J. B. 1989. Deconstructing the Map. *Cartographica*, 26: 1–20.

Harley, J. B. 1990. Cartography, Ethics and Social Theory. *Cartographica*, 27: 1–23.

Harley, J. B. 1991. Can There Be a Cartographic Ethics? *Cartographic Perspectives*, 10: 9–16.

Harley, J. B. 1992. Rereading the Maps of the Columbian Encounter. Annals, Association of American Geographers, 82(3): 522–42.

Harley, J. B. 2001. *The New Nature of Maps*. Baltimore, MD: Johns Hopkins University Press.

Harvey, D. 1996. *Justice, Nature and the Geography of Difference*. Oxford and New York: Blackwell Publishers.

Harvey, D. 2000. *Spaces of Hope*. Berkeley and Los Angeles: University of California Press.

Harvey, D. 2001. *Spaces of Capital: Towards a Critical Geography*. New York: Routledge.

Heidegger, M. 1962. *Being and Time*. Trans. J. Macquarrie and E. Robinson. Oxford: Basil Blackwell.

Heidegger, M. 1977. *The Question Concerning Technology and Other Essays*. Trans. W. Lovitt. New York: Harper Colophon Books.

Heidegger, M. 1993. Letter on Humanism. In D. F. Krell (ed.), *Martin Heidegger, Basic Writings*, 2nd edn. San Francisco: HarperSanFrancisco.

Heidegger, M. 1996. *Being and Time*. Trans. J. Stambaugh. Albany, NY: SUNY Press.

Heidegger, M. 2000. *Introduction to Metaphysics*. Trans. G. Fried and R. Polt. New Haven, CT and London: Yale University Press.

Hillner, J. 2000. Venture Capitals. *Wired*, July.

Hoffman, D. L. and Novak, T. P. 1998. Bridging the Digital Divide: the Impact of Race on Computer Access and Internet Use. *Science*, April 17.

Indymedia. 2001. Italian Cops Attempt Infiltration of Indymedia Chat Server. Online. http://sf.indymedia.org/display.php?id=102074

Ingraham, B. L. and Smith, G. W. 1972. The Use of Electronics in the Observation and Control of Human Behavior and Its Possible Use in Rehabilitation and Parole. *Issues in Criminology*, 7(2): 35–53.

Internet Software Consortium (ISC). 2002. Internet Domain Survey. Online. http://www.isc.org/

Johnson, R. W. 2002. Every Club in the Bag. *London Review of Books*, 8 August, pp. 15–16.

Kahney, L. 2001. Tracking Bloggers with Blogdex. *Wired*, July 31.

Kenyada, R. 2001. Personal Communication.

Kirsch, S. 1995. The Incredible Shrinking World? Technology and the Production of Space, *Environment and Planning D: Society and Space*, 13: 529–55.

KlaasKids Foundation. 2001. California Voters: You Can Help us Now! Online. http://web.archive.org/web/20010815180031/http://klaaskids.org/pg-dnalert.htm [Internet Archive].

Klein, N. 2000. *No Logo: Taking Aim at the Brand Bullies*. New York: Picador.

Klinkenberg, B. 1997. Unit 23 – History of GIS. NCGIA Core Curriculum. Online. http://www.geog.ubc.ca/courses/klink/gis.notes/ncgia/u23.html#UNIT23

Kominski, R. and Newburger, E. 1999. Access Denied: Changes in Computer Ownership and Use: 1984–1997. Paper Presented at the Annual Meeting of the American Sociological Association. Online. http://www.census.gov/population/socdemo/computer/confpap99.pdf

Konvitz, J. W. 1987. *Cartography in France 1660–1848. Science, Engineering, and Statecraft*. Chicago: University of Chicago Press.

Korzybski, A. 1948. *Science and Sanity: An Introduction to non-Aristotelian Systems and General Semantics*. 3rd edn. Lakeville, CT: Internal Non-Aristotelian Library.

Koskela, H. 2000. "The Gaze Without Eyes": Video-Surveillance and the Changing Nature of Urban Space, *Progress in Human Geography*, 24(2): 243–65.

Larrison, C. R., Nackerud, L., Risler, E. and Sullivan, M. Forthcoming. Welfare

Recipients and the Digital Divide: Left Out of the New Economy? Unpublished paper, University of Georgia.

Lasica, J. D. 2001. Why Are There Blogs? Online. http://jd.manilasites.com/2001/11/12#why

Lefebvre, H. 1991. *The Production of Space*. Oxford: Blackwell.

Liddell, H. G. and Scott, R. 1990 [1871]. *Greek-English Lexicon*. Abridged edn. Oxford: Oxford University Press.

Livingstone, D. N. 1992. *The Geographical Tradition, Episodes in the History of a Contested Enterprise*. Oxford: Blackwell.

Lotringer, S. (ed.). 1996. *Foucault Live, Interviews 1961–1984*. Trans. John Johnston. Expanded edn. New York: Semiotext(e).

Lyon, D. 1994. *The Electronic Eye. The Rise of the Surveillance Society*. Minneapolis, MN: University of Minnesota Press.

MacEachren, A. M. 1998. Cartography, GIS and the World Wide Web. *Progress in Human Geography* 22(4): 575–85.

McWhorter, L. 1999. *Bodies and Pleasures: Foucault and the Politics of Sexual Normalization*. Bloomington, IN: Indiana University Press.

Mark, D. M., Chrisman, N., Frank, A.U., McHaffie, P.H. and Pickles, J. Undated. The GIS History Project. Online. http://www.geog.buffalo.edu/ncgia/gishist/bar_harbor.html

Marsh, A. 2001. Mapping a pan-African Market. *Red Herring Magazine*, March 6. Online. http://www.redherring.com/mag/issue93/180018018.html

Mayor's Office of Community Technology (MOCT). 2002. About the Program. Online. http://www.atlantacommunitytech.com/about.htm

Mead, R. 2002. You've Got Blog. In Perseus Publishing (eds), *We've Got Blog, How Weblogs are Changing Our Culture*. Cambridge, MA: Perseus Publishing, pp. 47–56.

Minges, M. 2000. Counting the Net: Internet Access Indicators. Paper given at the 2000 INET Conference. Online. http://www.isoc.org/inet2000/cdproceedings/8e/8e_1.htm

Mishel, L., Bernstein, J. and Schmitt, J. 2001. *The State of Working America 2000–2001*. Ithaca, NY: ILR Press/Cornell University Press.

Monmonier, M. S. 1982. *Computer-Assisted Cartography, Principles and Prospects*. Englewood Cliffs, NJ: Prentice Hall.

Monmonier, M. S. 1985. *Technological Transition in Cartography*. Madison, WI: University of Wisconsin Press.

Monmonier, M. 1995. *Drawing the Line, Tales of Maps and Cartocontroversy*. New York: Henry Holt.

Monmonier, M. 1996. *How to Lie with Maps*, 2nd edn. Chicago: University of Chicago Press.

Monmonier, M. 1999. Epilogue. *Cartography and GIS*, 26(3): 235–6.

Monmonier, M. 2001. *Bushmanders and Bullwinkles. How Politicians Manipulate Electronic Maps and Census Data to Win Elections*. Chicago: University of Chicago Press.

Monmonier, M. 2002a. *Spying with Maps*. Chicago: University of Chicago Press.

Monmonier, M. 2002b. Maps, Politics and History. Interviewed by J. W. Crampton. *Environment and Planning: Society and Space* D, 20(6): 637–46.

National Telecommunications and Information Administration (NTIA) and Economics and Statistics Administration (ESA), 2000. *Falling Through the Net: Towards Digital Inclusion*. Washington, DC: Department of Commerce. Online. http://search.ntia.doc.gov/pdf/fttn00.pdf

National Telecommunications and Information Administration (NTIA) and Economics and Statistics Administration (ESA). 2002. *A Nation Online: How Americans are Expanding Their Use of the Internet*. Online. http://www.ntia.doc.gov/ntiahome/dn/anationonline2.pdf

Newburger, E. C. 1997. *Computer Use in the United States. Current Population Reports*, US Bureau of the Census.

Nielsen//Netratings. 2000. Internet Penetration Breaks The 50 Percent Mark In 21 U.S. Markets During September. Online. http://www.nielsen-netratings.com/pr/pr_001016.htm

Nietzsche, F. 1997. *Untimely Meditations*, ed. D. Breazeale. Cambridge: Cambridge University Press.

Nua.com. 2002. How many online? Online. http://www.nua.ie/surveys/how_many_online/

Olsson, G. 1998. Towards a Critique of Cartographical Reason. *Ethics, Place and Environment*, 1: pp. 145–55.

Olsson, G. 2002. Glimpses. In P. Gould and F. R. Pitts (eds), *Geographical Voices. Fourteen Autobiographical Essays*. Syracuse, NY: Syracuse University Press, pp. 237–68.

Oxfam International, 2001. Debt Relief. Still Failing the Poor. April. Online. http://www.oxfamamerica.org/news/art655.html

Paretsky, S. 1982. *Indemnity Only*. New York: Dell.

Patterson, T. 2002. *Getting Real: Reflecting on the New Look of National Park Service Maps*. Paper given at the 2002 Mountain Cartography Workshop, Mt Hood, Oregon.

Peet, R. 1998. *Modern Geographical Thought*. Oxford: Blackwell Publishers.

Peng, Z.-R. 1999. An Assessment of the Development Strategies of Internet GIS. *Environment and Planning B: Planning and Design*, 26(1): 117–32.

Petchenik, B. B. (ed.). 1988. Reflections on the Revolution: the Transition from Analogue to Digital Representations of Space, 1958–1988. Special Issue, *The American Cartographer*, 15(3).

Peterson, M. P. 1995. *Interactive and Animated Cartography*. Englewood Cliffs, NJ: Prentice-Hall.

Peterson, M. P. 1999a. Trends in Internet Map Use – A Second Look. In C. P. Keller (ed.), *Proceedings of the 19th International Cartographic Association*. Ottawa: ICA, pp. 571–80.

Peterson, M. 1999b. Maps on Stone: the Web and Ethics in Cartography. *Cartographic Perspectives*, 34: 5–8.

Philo, C. 1992. Foucault's Geography. *Environment and Planning D: Society and Space*, 10: 137–61. Reprinted in: M. Crang and N. Thrift (eds), *Thinking Space*. New York: Routledge.

Pickles, J. 1985. *Phenomenology, Science and Geography: Spatiality and the Human Sciences*. Cambridge: Cambridge University Press.

Pickles, J. 1991. Geography, GIS, and the Surveillant Society. *Papers and Proceedings of the Applied Geography Conference*, 14: 80–91.

Pickles, J. (ed.). 1995. *Ground Truth: The Social Implications of Geographic Information Systems*. New York: Guilford Press.

Pickles, J. 1997. Tool or Science? GIS, Technoscience and the Theoretical Turn. *Annals, Association of American Geographers*, 87(2): 363–72.

Pickles, J. 1999. Arguments, Debates and Dialogs: the GIS–Social Theory Debate and

the Concern for Alternatives. In P. A. Longley, M. F. Goodchild, D. J. Maguire and
D. W. Rhind (eds), *Geographical Information Systems*, 2nd edn. New York: John
Wiley, pp. 49–60.

Plewe, B. 1997. *GIS Online: Information Retrieval, Mapping, and the Internet*. Santa Fe,
NM: OnWord Press.

Polt, R. F. H. 1999. *Heidegger, an Introduction*. Ithaca, NY: Cornell University Press.

Porter, T. M. 1986. *The Rise of Statistical Thinking 1820–1900*. Princeton, NJ:
Princeton University Press.

Priest, C. 1999. *eXistenZ*. New York: HarperCollins.

Putz, S. 1994. Interactive Information Services Using World-Wide Web Hypertext.
Computer Networks and ISDN Systems, 27(2): 273–80.

Rabinow, P. (ed.). 1997. *Ethics Subjectivity and Truth. Essential Works of Michel Foucault
1954–1984, Vol. I*. New York: New Press.

Rheingold, H. 1993. *Virtual Community, Homesteading on the Electronic Frontier*.
Reading, MA: Addison-Wesley.

Robinson, A. H. 1971. The Genealogy of the Isopleth. *Cartographic Journal*, 8: 49–53.

Robinson, A. H. 1982. *Early Thematic Mapping in the History of Cartography*. Chicago:
University of Chicago.

Robinson, A. H. and Petchenik, B. B. 1976. *The Nature of Maps*. Chicago: University of
Chicago Press.

Rorty, R. 1999. Religion as Conversation-Stopper. In R. Rorty, *Philosophy and Social
Hope*. London and New York: Penguin Books, pp. 168–74.

Rose, N. 1996. *Inventing Our Selves. Psychology, Power, and Personhood*. Cambridge:
Cambridge University Press.

Rosen, J. 2001. A Watchful State. *New York Times Magazine*, October 7, pp. 38–43,
85–92.

Rossmo, D.K. 2000. *Geographic Profiling*. Boca Raton, FL: CRC Press.

Sallis, J. 1999. *Chorology. On Beginning in Plato's Timaeus*. Bloomington, IN: Indiana
University Press.

Scarborough Research. 2000. Personal Computer Ownership by DMA. Online.
http://www.scarborough.com/scarb2002/press/pr_hhpc_data.htm.

Schatzki, T. R. 1991. Spatial Ontology and Explanation. *Annals of the Association of
American Geographers*, 81(4): 650–70.

Schwartz, M. 1998. Critical Reproblemization. Foucault and the Task of Modern
Philosophy. *Radical Philosophy*, 91: 19–29.

Shenk, D. 1998. *Data Smog: Surviving the Information Glut*. San Francisco: Harper.

Simama, J. 2001. Race, Politics, and Pedagogy of New Media: From Civil Rights to
Cyber Rights. In J. T. Barber and A. A. Tait (eds), *The Information Society and the
Black Community*. Westport, CT: Praeger, pp. 193–214.

Sloterdijk, P. 1987. *Critique of Cynical Reason*. Minneapolis, MN: University of
Minnesota Press.

Smith, N. 1992. Real Wars, Theory Wars. *Progress in Human Geography*, 16(2):
257–71.

Spence, I. 2000. The invention and use of statistical charts. *Journal de la Société
Francaise de Statistique*, 141: 77–81.

Spence, I. and Wainer, H. 1997. William Playfair: A Daring Worthless Fellow. *Chance*,
10: 31–4.

Stalder, F. 2002. Privacy Is not the Antidote to Surveillance. *Surveillance and Society*, 1:
120–4.

Sullivan, A. 2002. Out of the Ashes, a New Way of Communicating. *Sunday Times*, February 24.

Sutin, L. 1991. *In Pursuit of VALIS: Selections from the Exegesis*. Novato, CA and Lancaster PA: Underwood-Miller.

Sutin, L. (ed.). 1995. *The Shifting Realities of Philip K. Dick. Selected Literary and Philosophical Writings*. New York: Vintage Books.

Taylor, P. J. 1992. Politics in Maps, Maps in Politics: A Tribute to Brian Harley. *Political Geography*, 11: 127–9.

Thrift, N. 2000. Entanglements of Power: Shadows? In J. P. Sharp, P. Routledge, C. Philo and R. Paddison (eds), *Entanglements of Power. Geographies of Domination/Resistance*. London and New York: Routledge.

Thrower, N. 1961. Animated Cartography. *Professional Geographer*, 11(6): 9–12.

Tobler, W. 1959. Automation and Cartography. *Geographical Review*, 49(4): 526–34.

Tobler, W. 1970. A Computer Movie Simulating Urban Growth in the Detroit Region. *Economic Geography*, 46(2): 234–40.

Torguson, J. S. 1997. User Interface Studies in the Virtual Map Environment. *Cartographic Perspectives*, 28: 29–31.

Tuathail, G. Ó. 1996. *Critical Geopolitics. The Politics of Writing Global Space*. Minneapolis, MN: University of Minnesota Press.

Tufte, E. R. 1983. *The Visual Display of Quantitative Information*. Cheshire, CT: Graphics Press.

Turkle, S. 1995. *Life on the Screen. Identity in the Age of the Internet*. New York: Simon & Schuster.

Tyner, J. 1992. *Introduction to Thematic Cartography*. Saddle River, NJ: Prentice-Hall.

United Nations Development Program (UNDP) 2001. *Human Development Report 2001*. New York and Oxford: Oxford University Press.

United States Census Bureau, 1997. NAICS 514191. Online Information Services. Online. http://www.census.gov/epcd/ec97/def/514191.HTM

United States Department of Commerce, National Telecommunications Information Administration (NTIA) and Economic and Statistics Administration (ESA). 2000. *Falling Through the Net: Toward Digital Inclusion. A Report on Americans' Access to Technology Tools*. Washington, DC.

United States Department of Commerce, National Telecommunications Information Administration (NTIA) and Economic and Statistics Administration (ESA). 2002. *A Nation Online: How Americans Are Expanding Their Use of the Internet*. Washington, DC.

United States Department of Education, National Center for Education Statistics (NCES). 2001. *Internet Access in US Public Schools and Classrooms: 1994–2000*. Online. http://nces.ed.gov/pubs2001/2001071.pdf

Unwin, T. 2000. A Waste of Space? Towards a Critique of the Social Production of Space. *Transactions of the Institute of British Geographers*, NS 25: 11–29.

Vatican Information Service (VIS). 2002. Internet: A New Forum for Proclaiming the Gospel. Online. http://www.vatican.va/holy_father/john_paul_ii/messages/communications/index.htm

Virilio, P. 1995. Red Alert in Cyberspace! *Radical Philosophy*, 74: 2–4.

Virilio, P. 1997. *Open Sky*. London: Verso.

Warner, M. 1999. *The Trouble with Normal. Sex, Politics, and the Ethics of Queer Life*. Boston, MA: Harvard University Press.

Warntz, W. 1983. Trajectories and Co-ordinates. In M. Billinge, D. Gregory and R.

Martin (eds), *Recollections of a Revolution, Geography as Spatial Science*. New York: St. Martin's Press.

Watson, H. (with the Yolngu community at Yirrkala). 1993. Aboriginal-Australian maps. In D. Turnbull, *Maps Are Territories. Science Is an Atlas*. Chicago: University of Chicago Press.

Weberman, D. 2000. Are Freedom and Anti-Humanism Compatible? The Case of Foucault and Butler. *Constellations, an International Journal of Critical and Democratic Theory*, 7(2): 255–71.

Weisburd, D. and McEwen, T. (eds). 1997. *Crime Mapping and Crime Prevention. Crime Prevention Studies*, Vol. 8. Monsey, NY: Criminal Justice Press.

Williams, J. 2002. Interview with John Ashcroft. September 11. National Public Radio. Online. http://search.npr.org/cf/cmn/segment_display.cfm?segID=149770

Wilson, M. I., Corey, K. E., Mickens, C. and Mickens, H. P. 2001. Death of Distance/Rise of Place: The Impact of the Internet on Locality and Spatial Organization. Paper prepared for INET 2001. Online. http://www.isoc.org/inet2001/CD_proceedings/U128/INET2001-U128.htm

Wireless NewsFactor. 2001. Digital Angel Takes Flight. Online. http://www.wirelessnewsfactor.com/perl/story/15059.html

Wood, D. 1987. Pleasure in the Idea: The Atlas as Narrative Form. *Cartographica*, 22: 24–45.

Wood, D. 1992. *The Power of Maps*. New York: Guilford Press.

Woodward, D. 1974. The Study of the History of Cartography: A Suggested Framework. *American Cartographer*, 1: 101–15.

Woodward, D. 2001a. "Theory" and *The History of Cartography*. In D. Woodward, C. Delano-Smith and C. D. K. Yee (eds), *Approaches and Challenges in a Worldwide History of Cartography*. Barcelona: Institut Cartogràfic de Catalunya.

Woodward, D. 2001b. The "Two Cultures" of Map History – Scientific and Humanistic Traditions: A Plea for Reintegration. In D. Woodward, C. Delano-Smith and C. D. K. Yee (eds), *Approaches and Challenges in a Worldwide History of Cartography*. Barcelona: Institut Cartogràfic de Catalunya.

World Bank. 2002. GNI per Capita 2001, Atlas Method. Online. http://www.worldbank.org/data/databytopic/GNIPC.pdf

Wright, D. J., Goodchild, M. F. and Proctor, J. D. 1997. GIS: Tool or Science? Demystifying the Persistent Ambiguity of GIS as "tool" versus "science". *Annals, Association of American Geographers*, 87(2): 346–62.

Yapa, L. S. 1991. Is GIS Appropriate Technology? *International Journal of Geographic Information Systems*, 5(1): 41–58.

Yapa, L. S. 1992. Why Do They Map GNP per Capita. In S. K. Majumdar, G. S. Forbes, E. W. Miller and R. F. Schmalz (eds), *Natural and Technological Disasters: Causes, Effects, and Preventive Measures*. Easton, PA: Pennsylvania Academy of Science.

Zakon, R. H. 2002. Hobbes' Internet Timeline. Online. http://www.zakon.org/robert/internet/timeline/

Index